NANOTECHNOLOGY SCIENCE AND TECHNOLOGY

DEVELOPMENT OF LUMINESCENCE PROPERTIES OF Eu3+-DOPED NANOSIZED MATERIALS

NANOTECHNOLOGY SCIENCE AND TECHNOLOGY

Additional books in this series can be found on Nova's website under the Series tab.

Additional E-books in this series can be found on Nova's website under the E-books tab.

NANOTECHNOLOGY SCIENCE AND TECHNOLOGY

DEVELOPMENT OF LUMINESCENCE PROPERTIES OF EU3+-DOPED NANOSIZED MATERIALS

LIXIN YU

Nova Science Publishers, Inc.
New York

Copyright © 2011 by Nova Science Publishers, Inc.

All rights reserved. No part of this book may be reproduced, stored in a retrieval system or transmitted in any form or by any means: electronic, electrostatic, magnetic tape, mechanical photocopying, recording or otherwise without the written permission of the Publisher.

For permission to use material from this book please contact us:
Telephone 631-231-7269; Fax 631-231-8175
Web Site: http://www.novapublishers.com

NOTICE TO THE READER

The Publisher has taken reasonable care in the preparation of this book, but makes no expressed or implied warranty of any kind and assumes no responsibility for any errors or omissions. No liability is assumed for incidental or consequential damages in connection with or arising out of information contained in this book. The Publisher shall not be liable for any special, consequential, or exemplary damages resulting, in whole or in part, from the readers' use of, or reliance upon, this material.

Independent verification should be sought for any data, advice or recommendations contained in this book. In addition, no responsibility is assumed by the publisher for any injury and/or damage to persons or property arising from any methods, products, instructions, ideas or otherwise contained in this publication.

This publication is designed to provide accurate and authoritative information with regard to the subject matter covered herein. It is sold with the clear understanding that the Publisher is not engaged in rendering legal or any other professional services. If legal or any other expert assistance is required, the services of a competent person should be sought. FROM A DECLARATION OF PARTICIPANTS JOINTLY ADOPTED BY A COMMITTEE OF THE AMERICAN BAR ASSOCIATION AND A COMMITTEE OF PUBLISHERS.

LIBRARY OF CONGRESS CATALOGING-IN-PUBLICATION DATA

Yu, Lixin, 1972-
  Development of luminescence properties of Eu3+-doped nanosized materials / author, Lixin Yu.
       p. cm.
   Includes bibliographical references and index.
   ISBN 978-1-61728-852-4 (softcover)
   1. Phosphors. 2. Nanocrystals--Materials. 3. Europium--Optical properties. 4. Luminescence. I. Title.
   QC476.7.Y83 2010
   620.1'1295--dc22
                                                             2010025952

Published by Nova Science Publishers, Inc. ✢ New York

CONTENTS

Preface		vii
Chapter 1	Introduction	1
Chapter 2	Light-Induced Changes of CT Band of Eu^{3+} Doped Nanocrystals	7
Chapter 3	Site Symmetry of Eu^{3+} Doped Nanocrysals	11
Chapter 4	The Dynamic Processes of Eu^{3+} Doped Nanocrystals	17
Chapter 5	Thermal Quenching Preoperties of Eu^{3+} Doped Nanocrystals	23
Chapter 6	The UCL of Eu^{3+} in Oxide Nanocrystal	27
Chapter 7	Spectral Hole Burning	31
Chapter 8	Core-Shell Nanopcomposites Doped with Eu^{3+}	33
Chapter 9	The Luminescence of Eu^{2+} in Nanocrystals	37
References		41
Index		53

PREFACE

It is well known that the reduction of nanocrytal size can result in the important modification of their properties in comparison with bulk materials because of the surface effect and size effect. Moreover, because Eu^{3+} ions are hypersensitive to the local microstructures, they can be used as fluorescent probe to search the local environment around Eu^{3+} ions. Thus Eu^{3+} doped nanomaterials exhibit unique luminescent properties, which are different from the micrometer hosts. In past decade, the luminescent characteristics of Eu^{3+} in nanomaterials have been extensively studied. In this review, we systematically discussed the recent development of photoluminescent properties of Eu^{3+} in different hosts, including the synthesis and morphology control (nanoparticles, nanowires/nanorods, nanopore and materials, etc) of nanomaterials (phosphate, borate, oxide, silicate etc) doped with Eu^{3+} prepared by different methods, surface state, the local symmetry surrounding Eu^{3+} ions, electronic transition processes, spectral changes of Eu^{3+} ions induced by light irradiation, upconversion luminescence, etc. The modification of luminescent properties of Eu^{3+} in nanocrystals in contrast with bulk hosts can help us develop advanced and high efficient nanophosphors.

Chapter 1

INTRODUCTION

The photoluminescence properties of rare-earth (lanthanide) compounds have been fascinating researches for decades [1-15]. An attractive feature of luminescent lanthanide compounds is their line-like emission, which results in a high color purity of the emitted light. The emission color depends on the lanthanide ion but is largely independent of the environment of a given lanthanide ion. For 17 rare earth (RE) elements (La, Ce, Pr, Nd, Pm, Sm, Eu, Gd, Tb, Dy, Ho, Er, Tm, Yb, Lu, Y and Sc), europium ions are important luminescent centers and appear trivalent and bivalent state (Eu^{3+} and Eu^{2+}). The emission of Eu^{3+} ($4f^6$) ions consists usually of some sharp lines in the red spectral area. These emissions have been found an important application in lighting and display devices. These emission lines correspond to transitions from the lowest excited level of $^5D_0$ to the ^7F_J (J=0-6) levels of the $4f^6$ configuration. Since the $^5D_0$ level will not be split by the crystal field, the splitting of the emission transitions yields the crystal field splitting of the ^7F_J levels. In addition to the emissions from $^5D_0$ levels, the emissions from higher ^5D levels, viz. $^5D_1$, $^5D_2$ and even $^5D_3$ are often observed. The $^5D_0$-^7F_J emissions are very suitable to survey the transition probabilities of the sharp spectral features of the RE ions. If a RE ion occupies in the crystal lattice site with inversion symmetry, optical transitions between levels of the $4f^n$ configuration are strictly forbidden as electric-dipole transition (parity selection rule). They can only take place as (the much weaker) magnetic-dipole transitions which obey the selection rule ΔJ=0, ±1 (but J=0 to J=0 forbidden) or as vibronic electric-dipole transitions. If there is no inversion symmetry at the site of the RE ion, the uneven crystal field components can mix opposite-parity states into the $4f^n$ configurational levels. The electric-dipole transitions are now no longer strictly forbidden and appear as (weak) lines in the spectra, the so-called forced electric-dipole transitions. Some transitions, viz. those with $\Delta J=0$, ± 2, are hypersensitive to this effect. Even for small deviations from inversion symmetry, they appear dominantly in the spectra. In fact, for applications it is required that the main emission is concentrated in the $^5D_0$-$^7F_2$ transitions (red emission). This illustrates the importance of hypersensitivity in materials research and spectroscopy [4].

Differently from trivalent Eu^{3+} ions, Eu^{2+} ($4f^7$) ions show a $5d$-$4f$ permitted transitions. The emission occurs broad band and varies from ultraviolet (UV) to yellow. Its decay time is about μs-order. This is due to the fact that the emitting level contains (spin) octets and sextets, whereas the ground state level (^8S from $4f^7$) is an octets, so that the spin selection rule slows down the optical transition rate. The host lattice dependence of emission color of Eu^{2+} ions is determined by the three factors: covalency; crystal field splitting of the $5d^1$ configuration; the Strokes shift. If the crystal field is weak and the amount of covalency is low, the lowest component of the $4f^65d$ configuration of the Eu^{2+} ions may shift to such high energy, that the $^6P_{7/2}$ level of the $4f^7$ configuration lies below it. At low temperatures, sharp-line emission due to the $^6P_{7/2}$-$^8S_{7/2}$ transitions occurs. Eu^{2+} ions activated compounds have been applied to the long-lasting phosphors [4].

The origin of zero-dimensional (0D) nanoparticles research can be from the study of the colloid solution, their synthesis and characteristics. Due to surface effect, quantum confinement effect of nanocrystals, the study of nanocrystals (including synthesis, physical and chemical properties) has been the locus since three to four decades [16]. In 1994, Bhargava *et al.* reported that the radiative transition rate of 0D ZnS: Mn^{2+} nanoparticles increased five-order in comparison with bulk one [17]. Despite this result was strongly criticized later, the studies on nanosized luminescent semiconductor attracted great interests, including the transition metal and RE ions-doped materials. In this review, we mainly concentrated on the Eu^{3+} doped nanocrystals. Up to now, some semiconductors nanocrystals doped with Eu^{3+} ions have been prepared and their luminescent properties were also studied, such as ZnS [18, 19], ZnO [20-23], TiO_2 [24-26], SnO_2 [27-29] and GaN [30, 31] nanocrystals, etc. Because there exist the large differences of radius between the trivalent RE ions and positive ions of semiconductor, the trivalent RE ions are difficult to enter into the lattice site of semiconductor nanocrystals, maybe locate at the surface. Moreover, it is difficult to prove the practical sites of Eu^{3+} in semiconductors nanocrystals. Recently, Chen et al. reported the site-selection spectra of Eu^{3+} in hexagonal ZnO nanocrystals prepared by a sol-gel method and Er^{3+} in TiO_2 nanocrystals [22, 32]. They proved that Eu^{3+} in ZnO and Er^{3+} in TiO_2 nanocrystals locate lattice sites through high resolution spectra at low temperature and theory calculation. Figure 1 show the site-selective spectroscopy of ZnO:Eu^{3+} at 10 K [22, 33]. It is clear that two kinds of luminescence sites of Eu^{3+} were identified. One site (B) exhibits a long lifetime of $^5D_0$ (1.16 ms) and sharp emission and excitation peaks, which is ascribed to inner lattice site with ordered crystalline environment. The other site (A) associated with distorted lattice sites near the surface shows a relatively short lifetime of $^5D_0$ (0.72 ms) and significantly broadened fluorescence lines. It is interesting that they observed the site conversion from A to B when the as-grown 4 nm nanocrystals were annealed at 400 °C for 1 h. Furthermore, as shown in figure 1 (3), typical Eu^{3+} luminescence from both sites can be seen under the band-gap excitation at 362 nm. The energy transfer (CT) from the nanocrystal host to Eu^{3+} confirms that Eu^{3+} ions can, to some extent, be incorporated into ZnO nanocrystals. The ET from ZnO host to Eu^{3+} will be more favorable because of the available matching levels of Eu^{3+} compared with the case of ZnS. Although some authors suggested that RE ion can enter into lattice of semiconductor nanocrystals, it should be noted that the luminescent intensity of RE ions in semiconductors is usually very low. In fact, the practical applications of RE doped phosphors hosts were normal lanthanide compound, shun as Y_2O_3, Gd_2O_3, YBO_3, YVO_4, etc. Thus in this paper, we mainly introduced the pholuminescence of Eu^{3+} doped lanthanide compounds nanocrystals. Lanthanide compounds nanocrystals doped with Eu^{3+} ions have been synthesized and their photoluminescent properties have been attracted considerable attentions because Eu^{3+} ions can be used as fluorescent probe to study the surface effect and local environments around RE ions in nanocrystals for decades. It is expected that nanocrystals doped with Eu^{3+} can improve the luminescent yield and display resolution. Thus different nanophosphors powders doped with Eu^{3+} with different shapes have been prepared. In earlier researches, most of topic was focused on the synthesis of nanocrystals doped with Eu^{3+} with different preparation techniques, such as solid state reactions, sol-gel techniques, hydroxide precipitation, hydrothermal synthesis, spray pyrolysis, laser-heated evaporation and combustion synthesis, et al. [34-38]. Some novle phenomena of Eu^{3+} ions in nanocrystals have been observed. Meltzer et al. studied the electron-phonon interaction in Eu_2O_3 nanoparticles by spectral burning hole and obtain the function of T^α ($3<\alpha<4$) between the homogeneous line width and temperature [39, 40]. They also reported that the radiative lifetime of Eu^{3+} in nanosized Y_2O_3 nanoparticles depended on the refractive index of the surrounding medium [40]. Several groups studied the local symmetry of Eu^{3+}-doped Y_2O_3 [41-44], YBO_3 [45-47], La_2O_3 [48] and $LaPO_4$ [49-51] nanocrystals (including nanoparticles and nanowires) through high resolution spectra at low temperature. The radiative lifetimes and dynamic processes of $^5D_0$-^7F_J and $^5D_1$-^7F_J of Eu^{3+} ions in nanocrystals were also researched [38, 43, 49-51, 52, 53]. It is interesting and important that the radiative transition rate and quantum efficiency (QE) of $^5D_0$ and $^5D_1$ level of Eu^{3+} in $LaPO_4$ nanowires were much higher than that in corresponding nanoparticles and bulk materials [49-51]. In addition, several groups found that the quenching concentration of Eu^{3+} ions in nanocrystals (nanopartilces and nanowires) is much higher than that in micrometer materials due to the hindrance of ET of the particle boundary [45, 46, 54, 55]. The similar phenomena were observed in Ce^{3+} or Tb^{3+} doped $LaPO_4$ nanowires [56, 57].

Because the disorder and proportion of surface atoms of nanocrystals increased in respect to the bulk materials, the nonradiative transition rate of RE ions in nanocrysatls was also increased. Recently the core-shell structures are of extensive scientific and technological interests. For photoluminescence of RE, we hope that the core-shell nanocomposites can reduce the nonradiative transition rate and improve the quantum yield. Moreover, the surface modification can achieve the functionalization of nanocrystals, such as biological compatibility which is permitted to be used as biological probe, chemical stabilities, etc. Accordingly, a large amounts of core-shell nanocomposites have been fabricated and the increase of luminescent intensity was observed, such as Y_2O_3:Eu^{3+}/SiO_2 [58], $CePO_4$:Tb^{3+}/$LaPO_4$ [59], $LaPO_4$:Eu^{3+}/$LaPO_4$ [60], YVO_4:Eu^{3+}/SiO_2 [61], Y_2O_3:Eu^{3+}/Y_2O_3 [62, 63], Y_2O_3:Eu^{3+}/SiO_2/YVO_4:Eu^{3+} [64], YBO_3:Eu/YVO_4 [65], and a serial of

SiO$_2$ coated with Eu^{3+}-doped nanophosphors by Lin et al. [66-72], and so on. These core-shell nanocomposites doped with RE can enlarge the application area and enrich the understanding of the optical and chemical properties of nanocrystals.

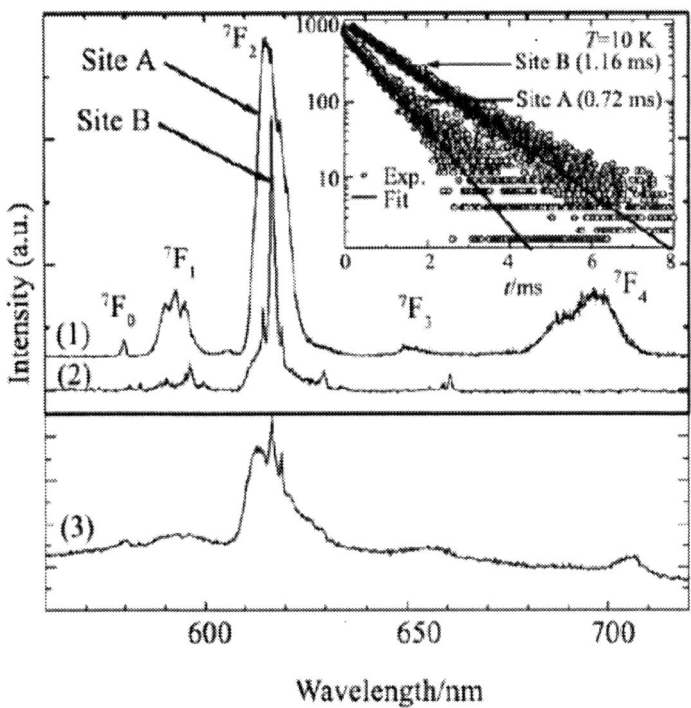

Figure 1. Site-selective emission spectra of Eu^{3+}:ZnO nanocrystals at 10 K, with (1) λ_{ex}=465.3 nm for site A, and (2) λ_{ex} =467.9 nm for site B. In (3) λ_{ex}, =362 nm, which corresponds to the band-gap excitation. The inset shows the 10 K decay curves from 5D$_0$ of Eu^{3+} at Sites A and B, respectively. Reprinted with permission from [22], Y. Liu et al. Opt. Lett. 2007, 32, 566.

In addition, some novel luminescent properties of Eu^{3+} ions doped nanocrystals were also reported. Song et al. reported that the changes of charge band (CT) band of Eu^{3+} doped Y$_2$O$_3$ nanocrystals with the irradiation by Xe-lamp [73]. Lu group investigated and compared the upconversion luminescence (UCL) properties of in Y$_2$O$_3$:Eu^{3+} nanocrtstals and Y$_2$O$_3$:Eu^{3+}@SiO$_2$ core-shell structures excited by the femtosecond (fs) laser [74].

In this book, we systematically discussed the recent development of photoluminescent properties of Eu^{3+} in different hosts, surface state, the local symmetry surrounding Eu^{3+} ions, electronic transition processes, spectral changes of Eu^{3+} ions induced by light irradiation, UCL, etc.

Chapter 2

LIGHT-INDUCED CHANGES OF CT BAND OF Eu^{3+} DOPED NANOCRYSTALS

In general, the trivalent ions that have a tendency to become divalent (Sm^{3+}, Eu^{3+}, Yb^{3+}) show CT absorption bands in the UV range. In the Eu^{3+}-doped oxide, electronic transition from $2p$ orbital of O^{2-} to the $4f$ orbital of Eu^{3+} produces excited CT band. The position of CT band of Eu^{3+} is a function of effective ionic radius of relative host lattice ion. The CT band appears in the spectra as broad absorption bands and generally locates at ~ 240 nm [75]. The Eu^{3+} doped nanocrystals, the location of CT reported by different authors was conflicting as a function of particle size and Eu^{3+} concentration. Tao *et al.* reported that the CT band red-shifted (longer wavelength) in $Y_2O_3:Eu^{3+}$ nanoparticles prepared by the combustion method when the particle size decreased from 80 to 5 nm [76]. Similar results were also observed in $Y_2O_3:Eu^{3+}$ nanoparticles [77, 78]. They attributed this phenomenon to the destabilization of O^{2-} in the nanocrystals making electron transfer occur at lower energy and longer Eu^{3+}-O^{2-} bond length in 8-fold coordination in the nanocrystals. On the contrary, Igarashi group reported that the CT band of Eu^{3+} in Y_2O_3 nanoparticles synthesized by the chemical vapor deposition blue-shifted in comparison with the bulk materials [79]. Konrad *et al.* also observed the same result in cubic $Y_2O_3:Eu^{3+}$ nanoparticles by the chemical vapor synthesis[80]. Fu *et al.* provided theoretical evidence of blue-shift of CT band of Eu^{3+} in SrY_2O_4 nanocrystals [81].

Song *et al.* reported that CT band of Eu^{3+}doped cubic Y_2O_3 nanoparticles by a combustion method did not shift with the variation of particle size [73]. But they found that the CT absorption decreased in intensity under Xe-lamp irradiation, while the bulk samples hardly changed after the same irradiation condition. Figure 2 shows the excited CT bands of nanocrystalline Y_2O_3 :Eu^{3+} (with the diameter of about 5 nm) before and after irradiation with various lights. The change was larger for smaller particle size and higher irradiation energy. They demonstrated that surface atoms enhances the proportion of local displacement for the nanoparticles with smaller particle size. The near surface of nanosized particle is particular unstable. Therefore, the local environments for the Eu^{3+} ions in the near surface are more easily rearranged by UV irradiation, leading to the change of CT band of Eu^{3+} after light irradiation. A special case is that the energy states of surface defects are close to $2p$ excited bands of O^{2-}. The electrons on the $2p$ excited states are captured by the nearby surface defects through tunneling, leading the CT states to decrease after UV irradiation. The essence of this process is still the optical rearrangements of local environments surrounding Eu^{3+} ions. Wang *et al.* systematically studied the changes of CT band with the irradiation condition, particles size and synthesis conditions [82].

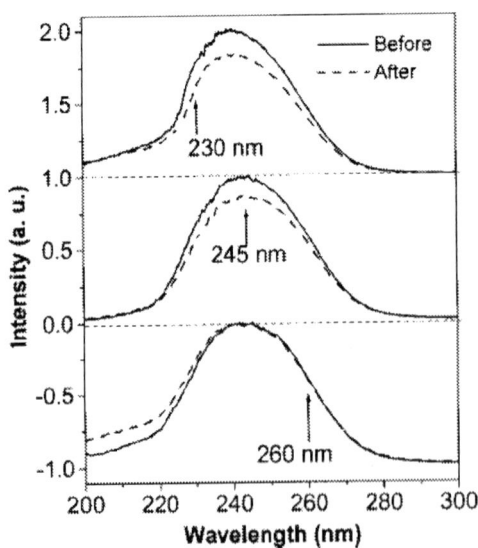

Figure 2. Excited CT bands of nanocrystalline Y_2O_3 :Eu^{3+}, measured before and after irradiation with various lights. Reprinted with permission from [73], H. Song *et al. Appl. Phys. Lett.* 2002, 81, 1776, American Institute of Physics.

After then, Bai et al. studied the changes of CT of $Y_2O_3:Eu^{3+}$ nanotubes synthesized by a hydrothermal method [43]. The outer diameters are of about 20-40 nm and length of ~1 μm. Figure 3 (a) shows the CT band before and after UV light irradiations with different wavelengths for 20 min in $Y_2O_3:Eu^{3+}$ nanotubes. It is obvious that the intensity in the CT band decreases as a whole as the sample is irradiated with a 230- or 245-nm light. As the sample is irradiated with a 260-nm light, the intensity at the longer wavelength side of the CT band has a small decrease, while that at the shorter wavelength side increases by a larger amount. This demonstrates that the UV irradiation induced spectral change is a function of the irradiation wavelength selective. The excited CT band before and after UV light irradiation were decomposed into two components also, as shown in figure 3 (b). Figure 3 (b) more obviously demonstrates that as the sample is irradiated with 230- or 246-nm light, the intensity in peak B decreases considerably, while that in peak A has only a small decrease. As the sample is irradiated with 260-nm light, the component of peak B decreases, while the component of peak A increases. It should be noted that the intensity variation in CT band under UV exposure depends strongly on the sample diameter. The present results are similar to that in $Y_2O_3:Eu^{3+}$ nanoparticles [73]. They attributed the present results to the local environment change surrounding Eu^{3+} ions, instead of the photoreduction from Eu^{3+} to Eu^{2+}.

In addition, several groups reported that the relative intensity of CT state of Eu^{3+} in nanocrystals with the variation of particle size and Eu^{3+} ions concentration. Pires et al. found that the relative intensity of CT band in $Gd_2O_3:Eu^{3+}$ (100-150 nm) increased with respective to the *f-f* transitions with increasing Eu^{3+} concentration [83]. Strek et al. reported that the CT band in $Lu_2O_3:Eu^{3+}$ nanoparticles increased in comparison with the *f-f* transitions absorption for larger sizes as well as for smaller concentration in the larger grains, and demonstrated that the surface population of Eu^{3+} ions was greater for the smaller particles [84]. In conclusion, the location of CT band depends on the Eu^{3+} concentration, the site symmetry and crystal lattice parameters etc [84]. In fact, the real content of Eu^{3+} in nanocrytsals lattice synthesized by different method with different sizes or morphologies was not further considered. The defects and microstructures around Eu^{3+} in nanocrystals by different synthesis techniques were possibly different. The conflicting results can help us to understand and consider these factors, such as the real concentration of Eu^{3+} ions, synthesis technique, shape effect and local sites of Eu^{3+} in naocrystals.

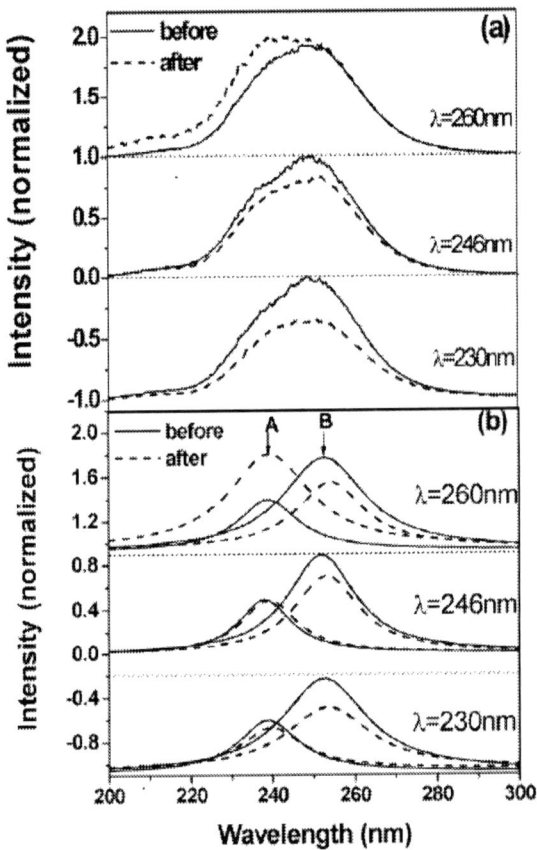

Figure 3. (a) Excited CT bands Eu^{3+} in Y_2O_3 nanotubes (λ_{em} = 611 nm) measured before and after irradiation with various lights. (b) Decomposed spectra. Reprinted with permission from [43], X. Bai *et al. J. Phys. Chem. B.* 2005, 109, 15236.

Chapter 3

SITE SYMMETRY OF Eu^{3+} DOPED NANOCRYSALS

In nanocrystals, a large number of atoms locate at the particle surface. Because of the existence of surface "dangling bonds" (i.e., coordination unsaturation) and "defect states/centers" (e.g., caused by the presence of impurities such as OH^- and CO_3^{2-}, etc), the local environments of the surface atoms/ions are naturally different from those of the inner ones. The authors of studies on crystalline LaF_3 nanoparticles doped with Eu^{3+} have established that the surface Eu^{3+} ions have a lower symmetry environment than those in the bulk [85]. The Eu^{3+} ion is an ideal fluorescent probes to detect local environments in nanocrystals due to the hypersensitivity of the $^5D_0-^7F_2$ transitions. In previous studies, researcheres found that the emission lines in nanocrystals become broader in comporison with those of the bulk as the result of more disordered local environments surrounding the luminescent centers [86]. Thus, the ratio of emission intensities, the red emissions of $^5D_0-^7F_2$ (R) to orange ones of $^5D_0-^7F_1$ (O) (R/O) was observed to increase with the decreasing particle size, implying that the chromaticity was improved with small particles size. Wei *et al.* observed the increased R/O ratio of $YBO_3:Eu^{3+}$ nanoparticles at vacuum ultraviolet excitation (147 nm) with the decrease of size, and results were listed in Table 1 [47]. Song *et al.* and Boyer *et al.* also observed the similar phenomena [50, 87]. The present results were attributed to the lower site symmetry of Eu^{3+} in nanocrystals.

Table 1. Optical characteristics of YBO$_3$:Eu$_{0.1}$ bulk and nanoparticles (a:17.5 nm, b:19.5 nm, c:28.1 nm, d: 32 nm, e: 40.4 nm, f: 90.2 nm, g:>500 nm) (Reprinted with permission from [47], Z. WWei et al. J. Appl. Phys. 2003, 93, 9783, American Institute of Physics

Samples	Relative luminescent intentisy		R/O in emission spectra	BA/CT in excitation spectra
	$^5D_0$-$^7F_1$ emission at 590 nm	$^5D_0$-$^7F_2$ emission at 615 nm		
a	0.015	0.072	2.761	
b	0.058	0.130	1.291	
c	0.203	0.333	0.946	0.769
d	0.279	0.427	0.883	0.860
e	0.368	0.537	0.843	0.912
f	0.742	0.890	0.693	0.978
g	1.945	1.898	0.563	1.426
bulk	1	1	0.577	1.420

In nanocrystalline samples of Y$_2$O$_3$:Eu^{3+} and CdWO$_4$:Eu^{3+}, the appearance of new emission lines was also reported [88, 89]. The appearance of new emission peaks may be mainly attributed to the appearance of additional phases and these cannot readily be identified by X-ray diffraction (XRD) patterns if present at concentrations of ca.<1%. Because there is no Stark splitting, the $^7F_0$-$^5D_0$ excitation spectra are very suitable to study the site symmetry. Song et al. studied the low temperature $^7F_0$-$^5D_0$ excitation spectra of Eu^{3+} in Y$_2$O$_3$ nanoparticles prepared by the combustion method, was shown in figure 4 [42]. They observed a broader excitation line at 579.9 nm, which accompanied the usual 580.6 nm line belongings to emission from Eu^{3+} at C$_2$ site symmetry. The relative intensity of the broad 579.6 nm peak was observed to depend strongly on particle size. In fact, figure 4 shows a stronger relative intensity contribution of the 579.6 nm band for smaller particle sizes. They also measured the site-selective luminescence dynamics, indicating that the lifetime of the emission from the peak at 579.6 nm (τ_s= 1.52 ms) was slightly shorter than from that at 580.6 nm (τ_s = 1.81 ms) [42], implying that the two excitation peaks were from the different sites. Bazzi et al. [90] studied the phase distribution and particle size of Y$_2$O$_3$:Eu^{3+} nanoparticles prepared by the gas-phase condensation and also observed similar phenomena for the excitation spectra of samples annealed at different temperatures. YBO$_3$:Eu^{3+} are extensively applied in vacuum UV displays. Pan et al. [91] preparred vaterite-type $YBO_3:Eu^{3+}$ nanocrystals with the hydrothermal method and study site-selection spectra, shown in figure 5. It can be seen that at least two excitation sites of $^7F_0-^5D_0$ were identified by the site-selective excitation spectra, whereas only one site was observed in the bulk material [91]. The peak is located at 580.84 nm (described as site A) and 579.48 nm (site B). The peak at 579.48 nm is broader than the one at 580.84 nm. The lifetimes of Eu^{3+} ions at the sites A and B were determined to be 0.59 ms and 2.69 ms, respectively, further proving that the two excitation peaks at 579.48 nm and 580.84 nm are from two crystalline sites. Dai et al. [92] studied the site-selection spectra of $ZnWO_4:Eu^{3+}$ nanocrystals synthesied a hydrothermal method and found the similar phenomina, shown in figure 6. It is obvious that in the monoclinic phase $ZnWO_4:Eu^{3+}$ nanocrystals, two excitation peaks of $^7F_0-^5D_0$, located at 579 nm (M) and 582 nm (N), were observed, as shown in figure 6. The band centered at 579 nm was relatively broader and its relative intensity contribution increased with decreasing particle size. In above the research, all authors suggest that the additional sites are all attributed to the surface effect because the additional sites in nanocrystals depend on the grain size.

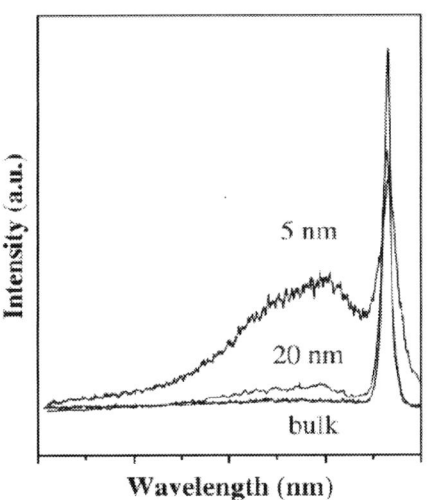

Figure 4. Excitation spectra of different size $Y_2O_3:Eu^{3+}$ powders at 10 K. The 580.6-nm peak intensity is normalized. Reprinted with permission from [42], H. Peng et al. Chem. Phys. Lett. 2003, 370, 485.

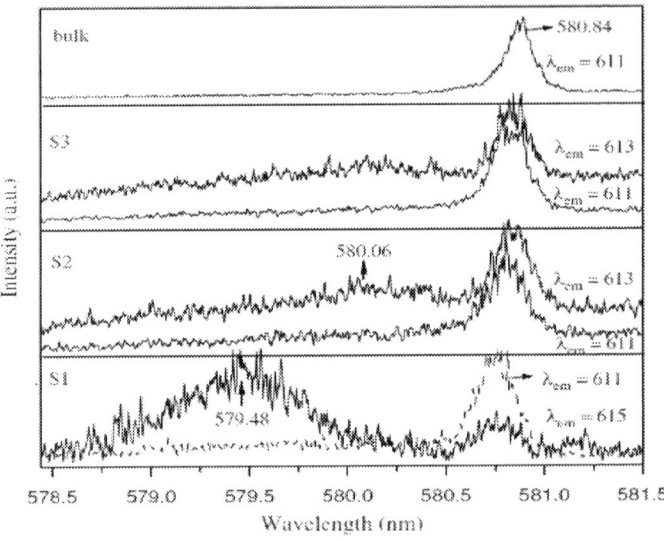

Figure 5. Site-selective excitation spectra of the $^7F_0$–$^5D_0$ transition of Eu^{3+} ions by monitoring different wavelengths (shown in nm) in YBO_3:Eu^{3+} nanocrystals. The samples a, b and c represent the hydrothermal products at 160, 180, and 200°C, respectively. Reprinted with permission from [91], G. Pan et al. J. Nanosci. Nanotech.. 2007, 7, 593.

In 2004, Yu et al. reported that the morphology effect affect the site symmetry of Eu^{3+} in $LaPO_4$ nanowires. Song and Yu et al. investigated the microstructures of Eu^{3+} doped one-dimensional (1D) $LaPO_4$ nanowires in contrast with 0D nanoparticles and corresponding micrometer hosts (micrometer particles and rods) [49-51]. The $LaPO_4$ nanoparticles, nanowires and bulk materiasl were synthesized by the same synthesis technique, a hydrothermal method. All samples were monazite phase and no additional phases were observed. The practical content of Eu^{3+} in $LaPO_4$ nano- and micro-crystals was nearly same (about 5%, mol ratio). The size of $LaPO_4$: Eu^{3+} nanoparticless is ranging from 10 to 20 nm. The diameter of $LaPO_4$: Eu^{3+} nanowires ranges from 10 to 20 nm, and the length ranges several hundreds nanometer. We observed that the Eu^{3+} ions locate new site in nanowires due to shape effect. Figure 7 shows the high resolution spectra of $LaPO_4$:Eu^{3+} nanomaterials and bulk materials at 266 nm pulsed laser excitation at 10 K. The emission associated with $^5D_0$-$^7F_1$ transitions is quite different between nanoparticles and nanowires at 10 K. In the nanoparticles, three

$^5D_0$-$^7F_1$ emission lines were observed, locating at 17025 ± 2 cm^{-1} (L1),16898 ± 2 cm^{-1} (L2) and 16815 ± 2 cm^{-1} (L3), respectively. In the nanowiress, besides the same lines L1-3, three additional lines L4-6 were observed, locating at 16963 ± 2 cm^{-1} (L4),16758 ± 2 cm^{-1} (L5) and 16718 ± 2 cm^{-1} (L6), respectively.

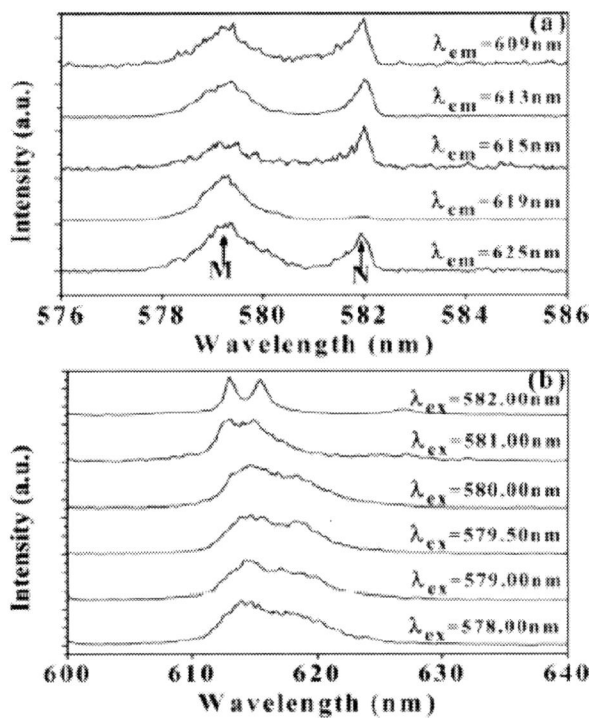

Figure 6. (a) Room temperature $^7F_0$–$^5D_0$ excitation spectra monitoring different $^5D_0$–$^7F_2$ bands, (b) $^5D_0$–$^7F_2$ emission spectra selectively exciting $^7F_0$–$^5D_0$ transitions in a sample of ZnWO$_4$:Eu^{3+} (2 mol%, prepared at 180°C, pH = 6). Reprinted with permission from [92], Q. Dai *et al. J. Phys. Chem .C* 2007, 111, 7586.

The $^5D_0$-$^7F_1$ lines in the microparticles were entirely identical with the nanoparticles. For the microrods, lines L1-6 were also observed. However, the relative intensity of lines L4-6 became weaker in comparison with that in nanowires. $^7F_1$ associated with one site symmetry can split into three Stark lines in the crystal field. The results in figure 7 indicated that in nanioparticles and microparticles, the $^5D_0$-$^7F_1$ transitions are from one crystalline site, A, while in nanowires and microrods, the $^5D_0$-$^7F_1$ transitions come from the same site (L1-L3), A, and an additional site (L4-L6), B. The relative number of Eu^{3+} at site B decreases as the powders varied from the nanowires to the microrods. In the present case, from the microparticles to the nanoparticles, the ratio of surface to volume increased greatly, but no additional site was observed. From the microrods to the nanowires, the ratio of surface to volume did not increase so much, however, the additional site B appeared and the relative number of Eu^{3+} at site B changed greatly. We thus believe that the appearance of the additional site B is not caused by the surface effect, but by the shape anisotropy. This is the first to report that the shape effect affects the local structures of Eu^{3+} doped 1D nanocrystals. This result can help me understand the nanosized effects on luminescent properties.

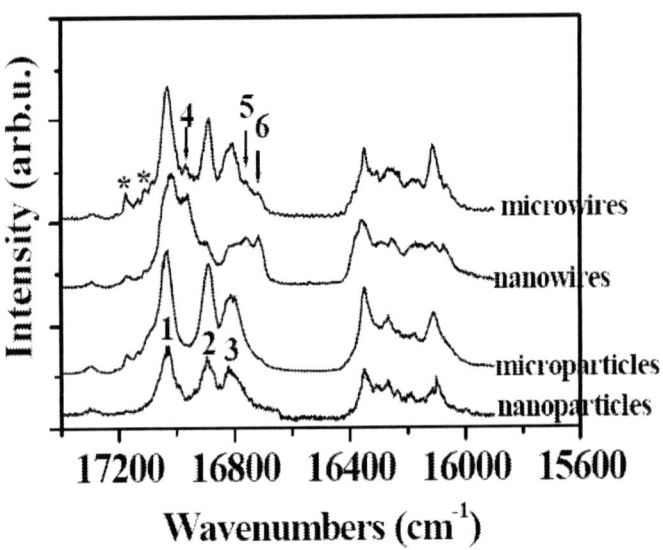

Figure 7. High-resolution spectra of $LaPO_4$:Eu^{3+} bulk, nanoparticless and nanowires at 10 K under 266 nm light excitation (delay time is 50 μs). Reprinted with permission from [49], L. Yu *et al. J. Phys. Chem .C.* 2004, 108, 16697.

Chapter 4

THE DYNAMIC PROCESSES OF Eu^{3+} DOPED NANOCRYSTALS

. The luminescent lifetime depends on the radiative and nonradiative decay rate. The converse of luminescent lifetime was proportional to quantum yield. The lifetime of RE depends on the host, microstructrues, medium, etc. Meltzer *et al.* reported the lifetime of the $^5D_0$ level of Eu^{3+} at the C-type site in monoclinic Y_2O_3 nanocrystals embedded in various matrices [40]. Two assumption were made: (1) the emission transitions from $^5D_0$ are electric dipole, and (2) the measured lifetime, τ, can be equated to the radiative lifetime, $τ_R$. And $τ_R$ can be expressed as:

$$\tau_R \times f(ED) \propto \frac{(\lambda_{vac})^2}{[n[(n^2+2)/3]^2]} \quad (1)$$

Where *f(ED)* is the oscillator strength of the electric dipole transition at vacuum wavelength λ, and *n* is the refractive index. The dependence of the radiative lifetime arises from the radiation field polarization of the medium and photon density change in an optically dense medium. The nanoparticles only occupy a small fraction of the volume of the medium, so that the effective refractive index $n(x)_{eff} = xn + (1-x)n_{med}$ is then more appropriate than *n*, where *x* is the filling factor that shows the fraction of the space occupied by the nanoparticles and n_{med} is the refractive index of the surrounding medium. Figure 8 shows the measured lifetimes for 12 nm Y_2O_3:Eu^{3+} nanoparticles embedded of dispersed in different media.

The three fitting lines was listed in figure 8 using Equ. 1. It can be seen that the experimental data were well fitted by Equ. 1. After then, several groups also studied the radiative lifetimes of $^5D_0$ of Eu^{3+} as the function of refractive index of media [93, 94]

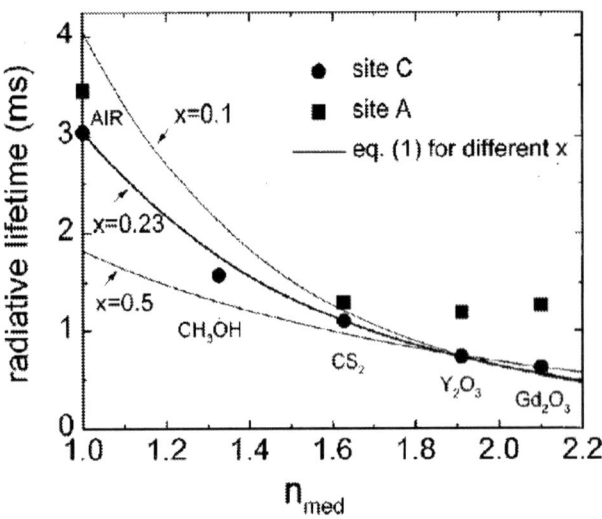

Figure 8. The dependence of the $^5D_0$ radiative lifetime, τ_R, for the Eu^{3+} C and A sites on the index of refraction of the media n_{med} at $T=295$ K. Solid lines—fits for nanocrystalline samples (site C) with Eq. (1) and different filling factors. Reprinted with permission from [40], R. Meltzer *et al. Phys. Rev. B* 1999, 60, R14012.

It is important if the quantum yield of $^5D_0$ or $^5D_1$ of Eu^{3+} in nanocrystal can be improved in contrast to the bulk materials. The luminescence radiative lifetime of $^5D_1$ level of Eu^{3+} is determined by the following equation:

$$\tau(T) = \frac{1}{W_1 + W_{10}(T)} \qquad (2)$$

where $\tau(T)$ is radiative lifetime of $^5D_1$, W_1 is the radiative transition rate of $^5D_1\text{-}\sum_J {}^7F_J$, $W_{10}(T)$ is nonradiative transition rate at a certain temperature, T. According to the theory of multi-phonon relaxation, W_{10} can be written as,

$$W_{10}(T) = W_{10}(0)(1+\langle n \rangle)^{\Delta E_{10}/\hbar\omega} \tag{3}$$

where $W_{10}(0)$ is nonradiative transition rate at 0 K, ΔE_{10} is the energy separation between $^5D_1$ and $^5D_0$, $\hbar\omega$ is the phonon energy, k is Boltzmann' constant and $\langle n \rangle = 1/(e^{\hbar\omega/kT}-1)$ is the phonon occupation number. According to Eq. 2-3, the lifetime of $^5D_1$ can be expressed as,

$$\tau = \frac{1}{W_1 + W_{10}(0)[1-\exp(-\hbar\omega/kT)]^{-\Delta E_{10}/\hbar\omega}} \tag{4}$$

According to Equ. 4, the radiative and nonradiative transition rate can be determined by fitting, if the lifetime of $^5D_1$ level of Eu^{3+} ions at different temperatures was measured. Song et al. studied the dynamic processes of Eu^{3+} doped cubic Y_2O_3 nanoparticles prepared by the combustion method [52]. Figure 9 is the dependence of fluorescent lifetime of $^5D_1$–$^7F_1$ on temperature. The fitted results using Euq. 4 were listed in Table 2. It appears that both radiative and nonradiative rates increase with decreasing particle size. In RE ions, the diameter of electronic wavefunctions of f states is in the order of 0.1 nm, which is much smaller than the particle size. In this case, the quantum confinement effect does not work. The increase of radiative transition rate with decreasing particle size is attributed to the lower local symmetry surrounding Eu^{3+}. In nanoparticles, more atoms are located in/near the particle surface, and numerous surface defects exist. These defects may increase the degree of disorder and lower the local symmetry of Eu^{3+} ions located in/near the surface of the particles. As a consequence, the radiative transition rate of $^5D_1$–^7F_J increases. Because the increase of nonradiative rate is more than the radiative rate, the QE of $^5D_1$ level of Eu^{3+} in Y_2O_3 nanoparticles decreased.

Table 2. Variation of radiative transition (W_1) and nonradiative relaxation rates (W_{10}) with particle size. Reprinted with permission from [52], H. Song et al. Chem. Phys. Lett. 2003, 376, 1

Particle size	5-nm	20-nm	3-im
W_1 (ms^{-1})	4.7	4.1	3.8
W_{10} (ms^{-1})	4.5	3.2	2.7

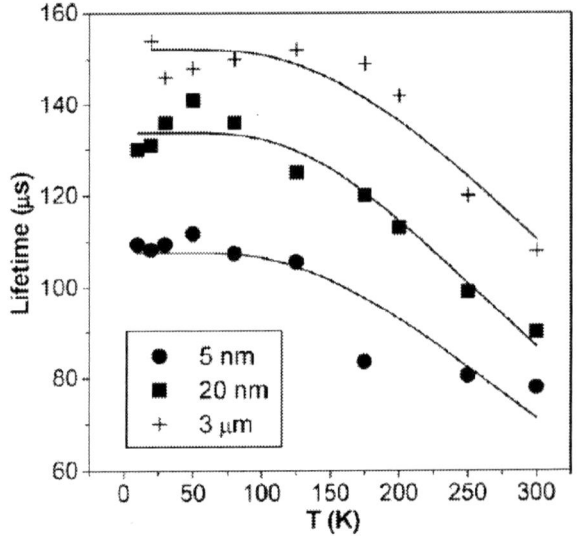

Figure 9. Dependence of fluorescent lifetime of $^5D_1-^7F_1$ on temperature. The dots are experimental data and the solid curves are fitting functions. A: nanoparticles (2 nm); B: bulk polycrystals. Reprinted with permission from [52], H. Song et al. Chem. Phys. Lett, 2003, 376, 1.

We also studied the electronic transition processes of Eu^{3+} in 1D $LaPO_4$ nanowires in comparison with 0D nanoparticle and bulk materials. It should be noted that all samples were prepared by the same method, viz. a hydrothermal method. Moreover, the Eu^{3+} extent in all samples is nearly same and the cross-relaxation among Eu^{3+} does not take place. The lifetime results were shown in figure 10. The fitted results using Equ. 4 were listed in Table 3. It can be seen that the radiative transition rate in the microsized materials (including micropaticles and microrods) was almost same. In comparison with the micrometer mateirals, the radiative transition rate in the nanoparticles decreased a little, while that in the nanowires greatly increased. In comparison with the bulk materials, the nonradiative transition rate of $^5D_1$-$^5D_0$ in nanopartices and nanowires had a little increase. The internal luminescent QE for $^5D_1$ level at 0 K, determined by $\eta = W_1 /[W_1 + W_{10}(0)]$ was also listed in Table 3. It is interesting and important that the QE of $^5D_1$ level in LaPO$_4$ nanowires is much higher than that in nanoparticles and bulk materials. For $^5D_0$ level, the same rules were obtained. The increased radiative transitive rate of $^5D_1$ and $^5D_0$ was attributed to the changes of polarization field of nanowires.

Table 3. A list of parameters W_1, $W_{10}(0)$ and the internal luminescent QE at 0 K in different powders. Reprinted with permission from [49], L. Yu et al. J. Phys. Chem .C. 2004, 108, 16697

Parameters	nanoparticles	nanowires	microparticles	microrods
W_1 (ms^{-1})	14.9	28.9	17.6	16.5
W_{10} (ms^{-1})	24.1	19.7	17.8	18.5
QE	38%	59%	49%	47%

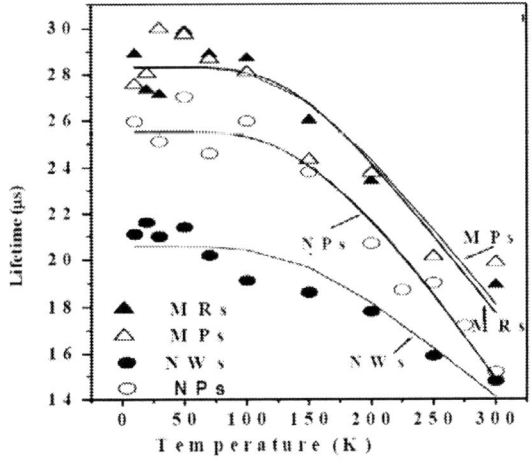

Figure 10. Dependence of emission lifetime of $^5D_1$ measured from the $^5D_1 - ^7F_2$ transitions upon temperature in various LaPO$_4$:Eu^{3+} samples: nanoparticles (NP); nanowires (NW); microparticles (MP); and microrods (MR). Scatter dots are experimental data and solid lines are fitting lines. Reprinted with permission from [49], L. Yu et al. J. Phys. Chem .C. 2004, 108, 16697.

Chapter 5

THERMAL QUENCHING PREOPERTIES OF Eu^{3+} DOPED NANOCRYSTALS

The thermal quenching characteristics of RE ions are impotant for high temperature application. Peng *et al.* studied the thermal quenching characteristics of Eu^{3+} in Y_2O_3 nanoparticles by a combustion method [53]. Figure 11 shows the temperature-dependent emission intensity of Eu^{3+} under the excitation of different lights (a) 580 nm; (b) 488 nm. Under the 580 nm excitation (the 580 nm light is in resonance with the $^7F_0$-$^5D_0$ transition), two main factors lead the emission intensity of Eu^{3+} to decrease with the elevated temperature. One factor is the thermal activated distribution of electrons among ^7F_J. Since the ground state $^7F_0$ is close to the other states $^7F_1$ and $^7F_2$, some electrons on $^7F_0$ will be thermally excited into $^7F_1$ and $^7F_2$ as temperature increases, causing the electron population on $^7F_0$ and the resonant transition of $^7F_0$ -$^5D_0$ to decrease. The other factor is the thermal quenching effect of luminescence, which is generally caused by nonradiative transition and ET processes. They constructed the equations of luminescent intensity as the function of temperature for $^5D_0$-$^7F_2$ transitions under 580 nm excitation (Equ. 5) and 488 nm excitation (Equ. 6)

$$I(T) = \frac{I_C \sigma_0 n_0(T)}{\{[\sum_j g_j (\exp(-\Delta E_{J0}/kT)(1+\beta \exp(T/T_C)]\}} \quad (5)$$

Where $n_0(T)$ is the population of $^7F_0$, σ_0 is the absorption cross section from $^7F_0 \nwarrow {}^5D_0$, I_C is the excitation intensity of the 580 nm light, $g_J=2J+1$ is the energy level degeneracy of ^7F_J (J=0, 1, 2), ΔE_{J0} is the energy separation from ^7F_J to $^7F_0$ (J=1, 2), k is Boltzmann's constant, $\beta=w_T(0)/w_{0R}$, $w_T(0)$ is the thermal quenching rate at 0 K, T_C is a temperature constant. w_{0R} is the radiative transition rate of $^5D_0$-^7F_J (J=0-2).

$$I(T) = \frac{I_C^{'} \sigma_2 n_0(0) g_2 \exp(-\Delta E_{20}/kT)}{\{[\sum_j g_j (\exp(-\Delta E_{J0}/kT)(1+\beta \exp(T/T_C)]\}} \quad (6)$$

where σ_2 is the absorption cross section from $^7F_2$-$^5D_2$, $I_C^{'}$ is the excitation intensity of the 488 nm light. Other parameters are the same with the Equ. 5. Equ. 5 and Equ. 6 were well fitted the experimental data. The fitting results were listed in Table 4. It is interesting that the values of parameter T_C were determined to be 310 in nanoparticles and 102 K in bulk host, respectively. This result means that the thermal quenching temperature in the nanoparticles is much higher than that in the bulk samples. They suggested that the stronger thermal quenching effect in nanocrystals was related to surface effect. Yu et al. also observed that the thermal quenching temperature of Eu^{3+} in $LaPO_4$ nanoparticles was much higher than that in bulk materials [95]. Song et al. reported that the similar phenomena in Y_2O_3:Tb^{3+} nanocarticles [96].

Table 4. A list of parameters w_{0R}, $w_{NR}(0)$ obtained by formula (6) in different Y_2O_3:Eu^{3+} powders. Reprinted with permission from [53], H. Peng et al. J. Chem. Phys. 2003, 118, 3277, American Institute of Physics

Parameters	sample A (4 nm)	Sample B (bulk polycrystals)
WOR (s-1)	536	923
WNR (s-1)	124	344
B	8.6	7.9

The above results indicates that the thermal quenching characteristics of some RE ions in nanocrystals are better than that in microsized materials, and more suitable for the high temperature applications.

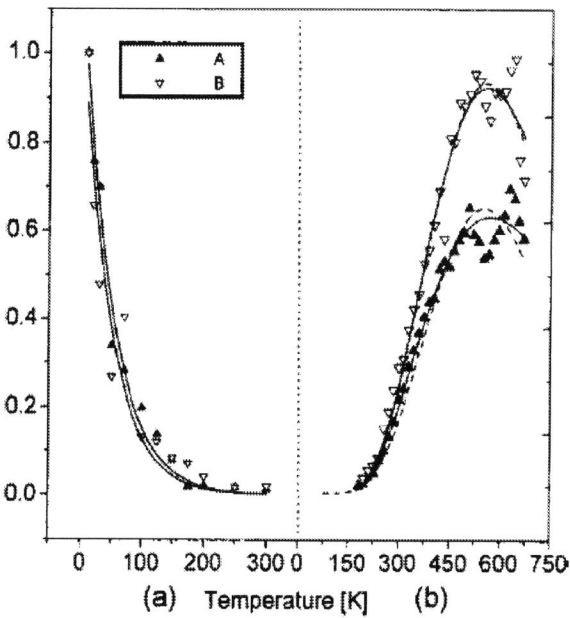

Figure 11. Temperature-dependent emission intensity of Eu^{3+} in Y_2O_3 nanoparticles under the excitation of different lights (a) 580 nm; (b) 488 nm. The dots are experimental data and the lines are fitting functions. Reprinted with permission from [53], H. Peng *et al. J. Chem. Phys.* 2003, 118, 3277, American Institute of Physics.

Chapter 6

THE UCL OF Eu^{3+} IN OXIDE NANOCRYSTAL

In previous, the studies of UCL were concentrated on the Er^{3+}, Tm^{3+}, Yb^{3+}, Ho^{3+}, Pr^{3+}, Sm^{3+} due to many fascinating applications in diode laser, solid-state-laser. The resorts on UCL of Eu^{3+} in crystallized and transparent hosts at near-infrared excitation are very rare since the energy levels of Eu^{3+} can not easily match with those of other RE ions. In fact, UCL of Eu^{3+} ions in different hosts can be achieved due to the advent of ultrafast titanium-sapphire tunable near-infrared laser that can produce fs laser and provide high-density to be used as excitation resource. As a new excitation source, the near-infrared fs laser has more advantages than other lasers. First, the fs laser can produce more intense pulses with the higher repetition rates and shorter pulse duration. Second, the higher excitation density of the fs pulse laser can achieve simultaneous multiphoton absorption more easily than the other pulse lasers. Recently the UCL of Eu^{3+}-doped Y_2O_3 nanoparticles and $Y_2O_3:Eu^{3+}/SiO_2$ core/shell nanostructures was reported by Lu *et al.* [74]. Figure 12 is the UCL spectra of $Y_2O_3:Eu^{3+}$ and $Y_2O_3:Eu^{3+}/SiO_2$ nanoparticles under fs laser irradiation (a), double logarithmic plots of UCL intensity of $Y_2O_3:Eu^{3+}$ nanoparticles as a function of pump power (b), and the UCL mechanisms (c). It is clear that the emissions of Eu^{3+} ions are enhanced after the nanoparticles are coated with SiO_2. The longer the coating time is, the stronger the intensity of Eu^{3+} ions is. In the surface of nanocrystals, a large number of surface defects exist, which act as the nonradiative transition channels, leading the nonradiative transition rates to increase and causing that luminescence becomes weak. In SiO_2-coated $Y_2O_3:Eu^{3+}$ nanocrystals, the surface defects are decreased because the cooperative ligand fields between the $Y_2O_3:Eu^{3+}$ core and non-crystalline SiO_2 shell interface active the 'dormant' Eu^{3+} ions near or on the surfaces of nanoparticles to participate in the UCL process, leading the intensity of Eu^{3+} in coated samples is stronger than that of noncoated samples.

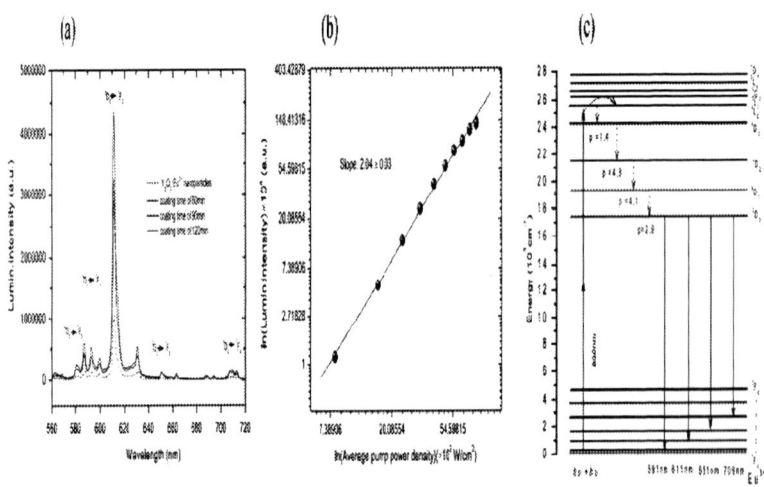

Figure 12. UCL spectra of $Y_2O_3:Eu^{3+}$ and $Y_2O_3:Eu^{3+}/SiO_2$ nanoparticles under 800-fs laser (a), double logarithmic plots of UCL intensity of $Y_2O_3:Eu^{3+}$ nanoparticles as a function of pump power (b), and the UCL mechanisms (c). Reprinted with permission from [74], Q. Lu *et al. Nanotech.* 2008, 19, 1.

Authors prove that the core-shell structures can increase the radiative quantum efficiency of Eu^{3+}-doped nanoparticles according to J-O theory analysis. In figure 12 (b), the slope of the double logarithmic fitted line is 2.04, indicating that the UCL of Eu^{3+} in Y_2O_3 nanocrystals is a two-photon process, which is different from the above mentioned UCL process of Eu^{3+} ions in silicate glasses [97, 98]. Authors proposed that the mechanism of UCL of Eu^{3+} ions is a simultaneous two-photon excitation process. The two-photon simultaneous absorption under 800-nm fs laser irradiation can be excited through a 'virtual' level to the excited level $^5L_6$ of Eu^{3+} with the aid of phonons of the host matrix. Detailed UCL process is shown in figure 12 (c).

As known, glass-ceramic consists noncrystalline glassy phase and nanocrystalline phase. Due to RE ions doped this material for their uses in telecommunication systems, such as upconversion fibers, optical amplifiers, solid-state lasers and 3D displays [99-101], efforts have been focus on finding materials with low-energy phonons, in order to reduce the multiphonon non-radiative de-excitation and improve cross-sections of the RE ions. In 2007, Qiu and coworker reported the UCL properties of Eu^{3+} in transparent SrO-TiO_2-SiO_2 glass-ceramics containing $Sr_2TiSi_2O_8$ microcrystals with a very large second-order optical nonlinearity for the first time [102]. Figure 13 show the emission spectra of Eu^{3+}-doped glass (a) and glass-ceramic (b) (left), the dependence of luminescent intensity of Eu^{3+}-doped glass (a) and glass-ceramic (b) on the fs laser pump power (right). The emission spectra of Eu^{3+} in both samples under 790-nm fs and 394-nm light excitation are similar. It should be noted that an emission at about 395 nm in the 790-nm fs laser irradiated glass-ceramic appears, while it is absent in the glass samples. This emission is assigned to the second-harmonic generation of the fs laser when excited by fs laser. It is vital important that the UCL intensity of Eu^{3+} ions in glass-ceramics is about 10 times stronger than that in the glass under the same condition because the absorption of the second-harmonic generation as a result of precipitated $Sr_2TiSi_2O_8$ microcrystalline particles with second-order optical nonlinearities at 790 nm fs laser excitation. The energy transfer from microcrystalites to Eu^{3+} ions takes place. The slope (n) of double logarithmic fitted lines is determined to be 1.90 in glass-ceramic and 1.5 in glass, indicating that the UCL process of Eu^{3+} ions in both glass-ceramics and glasses are two-photon simultaneous excitation at fs laser excitation (790 nm). Mostly important, the damage threshold of the glass-ceramics evidently increases in comparison with the glass. These results implied the glass-ceramics containing RE ions will be a promising three-dimensional display materials.

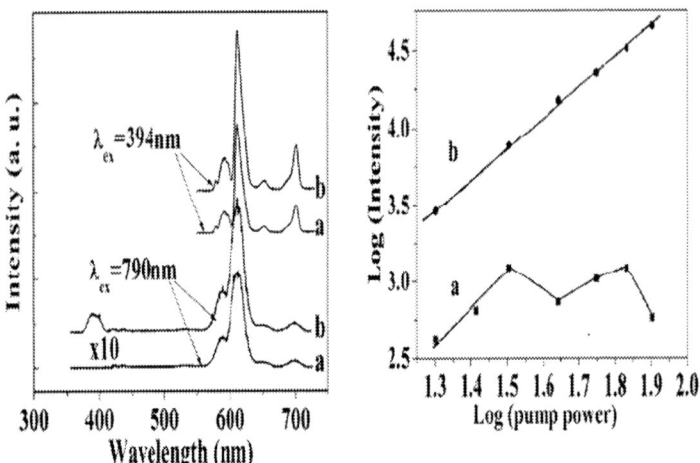

Figure 13. Emission spectra of Eu^{3+}-doped glass (a) and glass-ceramic (b) under 794-nm fs laser irradiation and 394-nm light excitation (left), the dependence of luminescent intensity of $^5D_0$-$^7F_2$ transitions of Eu^{3+}-doped glass (a) and glass-ceramic (b) on the fs laser pump power (right). Reprinted with permission from [102], B. Zhu *et al. Opt. Lett.* 2007, 32, 653.

Chapter 7

SPECTRAL HOLE BURNING

Spectral burning hole of RE ions has many potential applications in optical storages, optocommuniacation etc. Metltzer *et al.* studied the spectral burning hole performed on the $^5D_0$-$^7F_0$ transitions of Eu^{3+} in Eu_2O_3 nanocrystals [39]. The temperature dependences of the hole line widths are presented in figure 14. It is clear that the smaller nanoparticles have larger line widths at any given temperature than those of the larger particle samples.

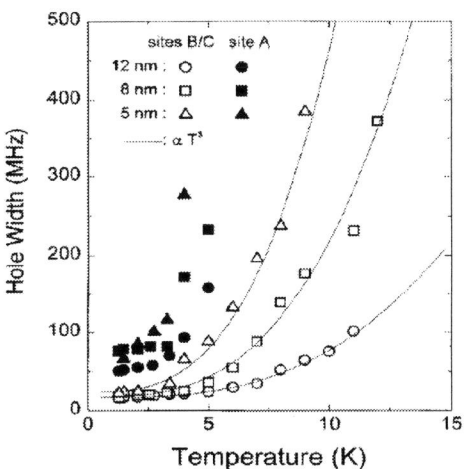

Figure 14. Hole width as a function of temperature for Eu_2O_3 nanoparticles with mean diameter of 5, 8 and 12 nm. The solid line is a fit to T^3. Reprinted with permission from [39], R. meltzer *et al. Phys. Rev. B.* 2000, 61, 3396.

The temperature dependence of the homogeneously broadened spectral linewidth of electronic transitions of lanthanide ions doped crystals and glasses depended on the interaction with phonons and depends on the phonon density [39]. It should be noted that they all have a line width which, at low temperatures, is nearly independent of temperature but which is much larger than the laser bandwidth limit (4 MHz). Above 4 K the line widths grow rapidly, approximately as T^a where $3<a<3.5$ for the B/C sites. Then several groups reported the vibrancy phonon spectra in Y_2O_3:Eu^{3+} and Gd_2O_3:Eu^{3+} nanoparticles [103-108].

Chapter 8

CORE-SHELL NANOPCOMPOSITES DOPED WITH Eu^{3+}

The core-shell structures can reduce the surface defects and achieve the surface modification. Earlier research was used to prepared the semiconductors nanocrystals. Accordingly, a large amounts of core-shell nanocomposites doped with RE ions have been fabricated and the increase of luminescent intensity was observed. In 1998, Li *et al.* prepared Y_2O_3:Eu^{3+} nanoparticles coated with SiO_2 or Al_2O_3 with sol-gel method and observed the considerable enhancement of luminescent intensity [58]. M. Haase reported the $CePO_4$:Tb^{3+}/$LaPO_4$ core-shelll colloidal resolution and the quantum yield of Tb^{3+} was increased from 43% to 70% [58]. This group also studied the $LaPO_4$:Eu^{3+}/ $LaPO_4$ core-shell nanoparticles and prove the surface sites of Eu^{3+} disappeared after surface coating [60]. LaF_3/Ln and $LaPO_4$/Ln core-shell nanocomposites were also reported. [109]. Bai *et al.* prepared the Y_2O_3:Eu^{3+} nanotubes coated with Y_2O_3 [62]. Figure 15 is excitation spectra of Y_2O_3:Eu^{3+} nanotubes and Y_2O_3:Eu^{3+} nanotubes coated with Y_2O_3 (a), emission spectra (b), the fluorescence decay curves of $^5D_0$ state (c) and the fluorescence decay curves of $^5D_1$ state (d). It can be seen that the luminescent intensity of Eu^{3+} in coated nanotubes is higher than that in noncoated samples. In the Y_2O_3:Eu^{3+} nanotubes, the $^5D_1$-$^7F_1$ transitions decay nonexponentially and much faster than that in the core-shell structure. The two decay time constants are determined to be τ_1=65.9 μs (with an intensity ratio of 49.6%) and τ_2=4.8 μs (50.4%) by a biexponential fitting. In the core-shell structure, the decay curve of $^5D_1$-

$^7F_1$ almost goes with a single exponential line, with a time constant of 94.6 μs. In contrast, the nonexponential decay in the nanotubes can be attributed to the ET from the excited state of Eu^{3+} ions to surface defect states, which should have a much larger rate than radiative transitions [110,]. For the $^5D_0$-$^7F_2$ transitions, the fluorescence decay is single exponential for both the samples, with a time constant of 1.7 ms in the nanotubess and of 2.1 ms in the core-shell structure. The above results demonstrate that the coating can reduce the nonradiative relaxation of the *f-f* inner-shell transitions for Eu^{3+} considerably, especially for the higher excited states. The core-shell composites also can be dedicated to realize novel or improved functions through special ET routes. YBO_3:Eu^{3+} is of vacuum ultraviolet luminescent hosts with high QE [111-115]. But because its characteristic emissions comprise almost equal contributions from $^5D_0$-$^7F_1$ (orange) and $^5D_0$-$^7F_2$ (red) transitions, its chromaticity is poor. YVO_4:Eu^{3+} with zircon structure is another important red phosphor. Because of the superior chromaticity, the strong $^5D_0$-$^7F_2$ transitions and the efficient ET from VO_4^{3-} to Eu^{3+}, the bulk YVO_4: Eu^{3+} phosphor with a quantum yield of 70% has been used as a red phosphor in cathode ray tubes for more than decades [116–119]. Recently Pan *et al.* prepared YBO_3:Eu^{3+}/ YVO_4:Eu^{3+} core-shell composites [65]. In this system, an effective ET route via YBO_3 phase→YVO_4 phase →Eu^{3+} ions was observed. As shown in figure 16, with excitation into VO_4^{3-} at 280 nm, intense red emissions, associated with the $^5D_0$-$^7F_2$ transition for Eu^{3+} ions distributed in the YVO_4 phase, dominate the emission spectrum. However, again for this system (S2), for excitation into the CT band in YBO_3:Eu^{3+} (240 nm), the orange–red emissions (with the strongest line being $^5D_0$-$^7F_1$ at 593 nm and with relatively weak $^5D_0$-$^7F_2$ lines at 611 and 627 nm) are not as prominent as in YBO_3:Eu^{3+} nanocrystals (S1). The red emissions of YVO_4:Eu^{3+} are strong so that the color purity is therefore improved remarkably in the YBO_3: Eu^{3+}/YVO_4:Eu^{3+} core-shell composites. Thereby, in these core-shell composites, the color purity was improved considerably, and the high quantum yield was maintained. These novel properties encourage the core-shell composites to be ideal candidate phosphors for red plasma display panels and Hg-free fluorescent lamps in future. The local environments of Eu^{3+} ions in the core-shell composites were also studied by high-resolution selective excitation spectra [65]. The results indicated that there exist interface symmetry sites and the local environmnts surrounding these interfaces become more complex. Figure 17 (left hand side) shows the $^7F_0$–$^5D_0$ site-selective excitation spectra in the composite (bottom spectrum) in contrast to the pure $YBO_3:Eu^{3+}$ nanocrystals (top spectrum). In the latter, two excitation lines (located at 580.8 nm (site 1) and 579.5 nm (site 2) were identified when monitoring two different emission wavelengths. These lines were assigned to the transitions of Eu^{3+} at internal and surface sites, respectively. In the composite, additional excitation lines also appeared and were assigned to the emissions of Eu^{3+} in the YVO_4 phase [65]. In comparison with the excitation line at the surface site in the $YBO_3:Eu^{3+}$ nanocrystals, the line at the interface site of the composite became broader as the result of more disordered local environments. The frequency-selective emission spectra indicated that unlike in the $YBO_3:Eu^{3+}$ nanocrystals, when pumping the Eu^{3+} ions located at the interface sites with different excitation energies, the emission spectra varied significantly. This was interpreted to imply complicated local environments of Eu^{3+} ions at the interface site [65].

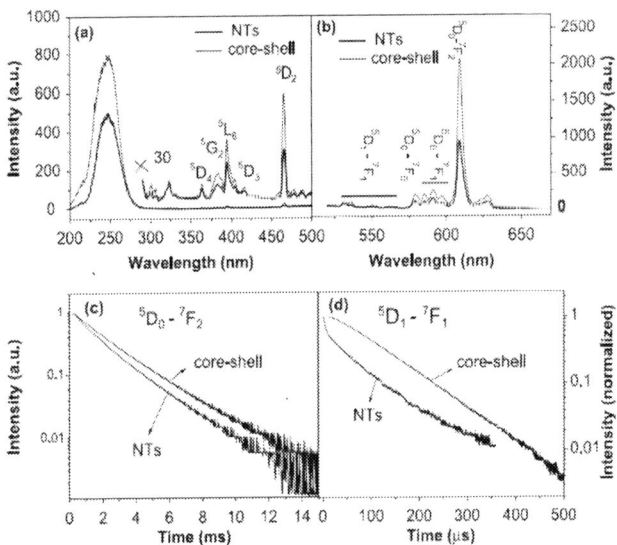

Figure 15. (a) Excitation spectra of $Y_2O_3:Eu^{3+}$ nanotubes and $Y_2O_3:Eu^{3+}@Y_2O_3$ (λ_{em}=611 nm), (b) emission spectra (λ_{ex} =246 nm), (c) the fluorescence decay curves of $^5D_0$ state (λ_{em}=611 nm) and (d) the fluorescence decay curves of $^5D_1$ state (λ_{em}=537 nm). Reprinted with permission from [62], X. Bai *et al. Appl. Phys. Lett.* 2006, 88, 143104, American Institute of Physics.

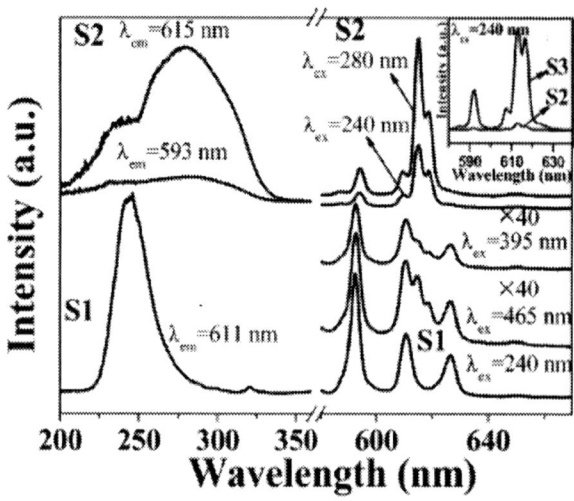

Figure 16. Excitation (left side) and emission (right-hand side) spectra of sample S2 (YBO$_3$:Eu^{3+}/YVO$_4$:Eu^{3+} core-shell composite) in contrast with sample S1 (YBO$_3$:Eu^{3+} nanocrystals). The inset is comparison of emission spectra of sample S2 with S3 (S3 is the annealed S2 sample). Reprinted with permission from [65], G. Pan *et al. Chem. Mater.* 2006, 18, 4526.

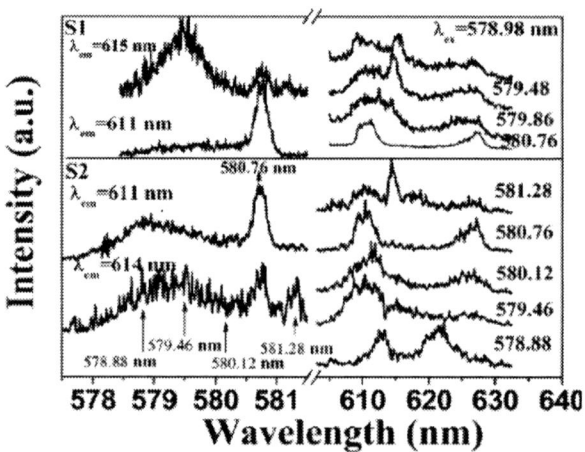

Figure 17. Comparison of site-selective excitation (left side) and emission spectra (right-hand side) of sample S2 (YBO$_3$:Eu^{3+}/YVO$_4$:Eu^{3+} core-shell composite) with sample S1 (YBO$_3$:Eu^{3+} nanocrystals). Reprinted with permission from [65], G. Pan *et al. Chem. Mater.* 2006, 18, 4526.

Chapter 9

THE LUMINESCENCE OF Eu^{2+} IN NANOCRYSTALS

Differently from Eu^{3+}, the *f-d* transitions of Eu^{2+} is permitted electronic dipole and appear broad band [4]. In nanocrystals, the luminescent properties of Eu^{2+} was also attracted considerable interests. In 2001, Chen *et al.* studied the luminescence of Eu^{2+} in ZnS nanoparticles [120]. Figure 18 shows the photoluminescence emission spectra of ZnS:Eu^{2+} nanoparticles with average sizes of 2.6, 3.2, and 4.2 nm, respectively. The emission band is from Eu^{2+} ions. It can be seen that the Eu^{2+} emission shifts to higher energies for smaller particles. It was demonstrated that quantum confinement effect can modify the energy structure of the nanocrystal host as well as the relative energy levels of the dopants. Specifically, the excited states of Eu^{2+} fall into the band gap of ZnS nanocrystals by size controlling, which thus enables the intra-ion transition of Eu^{2+} in ZnS nanocrystals. By contrast, no intra-ion transition can be observed in Eu^{2+}:ZnS bulk because of photoionization. F. Veggle *et al.* reported the photoluminescent characteristics of Eu^{2+} in GaN and GaN/SiO_2 nanocrystals. Figure 19 is the emission spectra of Eu^{2+} in GaN/SiO_2 nanocrystals. They proved that the Eu^{2+} was bivalent through ESR signals. The strong emission from the Eu^{2+} ions is assigned to the electronic transition occurring between the $4f^6$--$5d^1$ excited state and $^8S_{7/2}$ ground state of the Eu^{2+}. This transition is an allowed transition due to mixing of $5d$ states with $4f$ states in contrast to the parity forbidden transitions within the $4f$ energy levels of the lanthanides. The excited state configuration $4f^6-5d^1$ is very sensitive to the host lattice and can occur in any part of the visible region of the electromagnetic spectrum.

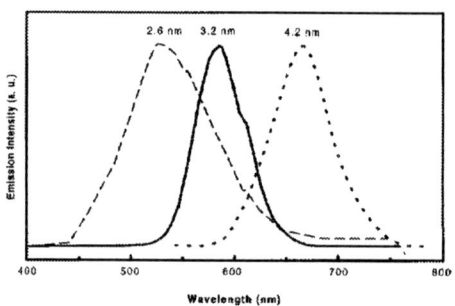

Figure 18. Fluorescence emission spectra of ZnS:Eu^{2+} nanoparticles with average size of around 4.2, 3.2, and 2.6 nm, respectively. The excitation wavelength is 260 nm. Reprinted with permission from [120], W. Chen *et al. J. Appl. Phys.* 2001, 89, 2671, American Institute of Physics.

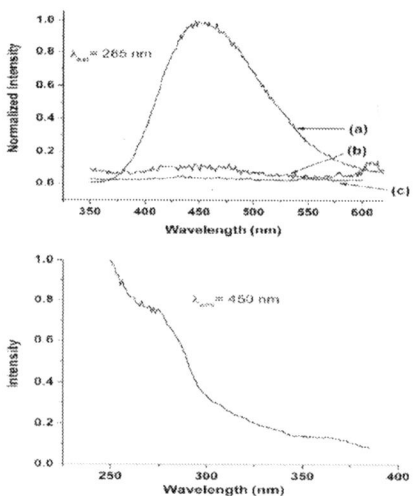

Figure 19. Top: PL emission spectra collected from a) Eu^{2+}-doped GaN/SiO$_2$ nanocomposites, b) Eu^{3+}-doped Ga$_2$O$_3$@SiO$_2$ after heated in air at 900 °C and c) bare SiO$_2$ nanoparticles after heated at 900 °C in NH$_3$ (ëex = 285 nm). Bottom: the excitation spectrum collected from the nanocomposites (ëex = 450 nm). Reprinted with permission from [30], V. Mahalingam *et al. Adv. Funct. Mater.* 2007, 17, 3462.

In conclusion, we systematically produce the progress of the photoluminescent properties of europium ions (Eu^{3+} and Eu^{2+}) in nanocrystal, including the changes of CT band of Eu^{3+} induced by light irradiation, local symmetry, dynamics processes, spectral burning hole, core-shell structures, etc.

REFERENCES

[1] K. Binnemans, Lanthanide-Based Luminescent Hybrid Materials. *Chem. Rev. 2009*, 109, 4283-4374.

[2] J.-C. G Bunzli, Luminescent probes. In lanthanide probes in life, chemical and earth sciences: theory and practice; Elsevier: Amsterdam, 1989.

[3] A. Kenyon, Recent developments in rare-earth doped materials for optoelectronics. *Prog. Quantum Electron.* 2002, 26, 225-284.

[4] G. Blasse, B. C.Grabmaier, Luminescent Materials; Spinger-Verlag: Berlin, 1994.

[5] G. Blasse, Luminescence of inorganic solids from isolated centers to concentrated systems. *Prog. Solid State Chem.* 1988, 18, 79-171.

[6] M. Elbanowski,B. Makowsaka, The lanthanides as luminescent probes in investigations biochemical systems. J. Photochem. Photobiol. A 1996, 99, 85-92.

[7] J. C. G. Bunzli, Benefiting from the unique properties of lanthanide ions. *Acc. Chem. Res.* 2006, 39, 53-61.

[8] J.-C G. Bunzli, C. Piguet, Taking advantage of luminescent lanthanide ions. *Chem. Soc. Rev.* 2005, 34, 1048-1077.

[9] Y. Hasegawa, Y. Wada, S. Yanagida, *J. Photochem. Photobiol. C* 2004, 5, 183-202.

[10] B. M. Tissue, Synthesis and luminescence of lanthanide ions in nanoscale insulating hosts. *Chem. Mater.* 1998, 10, 2837-2845.

[11] W. T. Carnall, J. V. Beitz, H. Crosswhite, K. Rajnak, J. B. Mann, Spectroscopic properties of the f-elements in compounds and solutions. In systematics and the properties of panthanides; *Sinha,*

S. P., Ed.; D. Reidel Publishing Company: Dordrecht, The Netherlands, 1983; p 389.

[12] W. T. Carnall, The absorption and fluorescence spectra of rare earth ions in solution. In handbook on the physics and chemistry of rare earths; *Elsevier: Amsterdam,* 1979; Vol. 3, Chapter 24, p 171.

[13] J. Blasse, Chemistry and physics of R-activated phosphors. In handbook on the physics and chemistry of rare earths; *Gschneidner, Elsevier: Amsterdam,* 1979; Vol. 4, Chapter 34, p 237.

[14] M. J. Weber, Rare earth lasers. In handbook on the physics and chemistry of rare earths; *Gschneidner, Elsevier: Amsterdam,* 1979; Vol. 4, Chapter 35, p 275.

[15] C. A. Morrison, R. P. Leavitt, Spectroscopic properties of triply ionized lanthanides in transparent host crystals. In handbook on the physics and chemistry of rare earths. *Elsevier: Amsterdam*, 1982; Vol. 5, Chapter 46, p 461.

[16] H. Chander, Development of nanophosphors-A review. *Mater. Sci. Engine. R* 2005, 49, 113-155.

[17] R. N. Bhargava, D. Gallagher, X. Kong, A. Nurmikko, Optical properties of manganese-doped nanocrystals of ZnS. *Phys. Rev. Lett.* 1994, 72, 416-419.

[18] S. C. Qu, W. H. Zhou, F. Q. Liu, N. F. Chen, Z. G. Wang, H. Y. Pan, D. P. Yu, Photoluminescence properties of Eu^{3+}-doped ZnS nanocrystals prepared in a water-methanol solution. *Appl. Phys. Lett.* 2002, 80, 3605-3607.

[19] W. Chen, J.e Malm, J. Bovin, Upconversion luminescence of Eu^{3+} and Mn^{2+} in $ZnS:Mn^{2+}$, Eu^{3+} codoped nanoparticles. *J. Appl. Phys.* 2004, 95, 667-673.

[20] A. Ishizumia, Y. Taguchia, A. Yamamotoa, Y. Kanemitsu, Luminescence properties of ZnO and Eu^{3+}-doped ZnO nanorods. *Thin Solid Films*, 2005, 486, 50–52.

[21] Y. Liu, W. Luo, R. Li, G. Liu, M. R. Antonio, X. Chen, Optical spectroscopy of Eu^{3+} doped ZnO nanocrystals. *J. Phys. Chem. C* 2008, 112, 686-694

[22] Y. Liu, W. Luo, R. Li, X. Chen, Spectroscopic evidence of the multiple-site structure of Eu^{3+} ions incorporated in ZnO nanocrystals. *Opt. Lett.* 2007, 32, 566-568.

[23] L. Armelao, F. Heigl, A. Julrgensen, R. I. R. Blyth, T. Regier, X.-T. Zhou, T. K. Sham, X-ray excited optical luminescence studies of ZnO and Eu-doped ZnO nanostructures, *J. Phys. Chem. C* 2007, 111, 10194-10200.

[24] Jan W. Stouwdam, Frank C. J. M. van Veggel, Sensitized emission in Ln^{3+}-doped TiO_2 semiconductor nanoparticles. *ChemPhysChem* 2004, 5, 743 -746.

[25] B. Julian, R.Corberan, E. Cordoncillo, P. Escribano, B. Viana , C. Sanchez, One-pot synthesis and optical properties of Eu^{3+}-doped nanocrystalline TiO_2 and ZrO_2. *Nanotech.* 2005, 16, 2707–2713.

[26] L. Li, C. Tsung, Z. Yang, G. Stucky, L. Sun, J. Wang, C. Yan, Rare-earth-doped nancocrystalline titania micropheres emitting luminescence via energy transfer. *Adv. Mater.* 2008, 20, 903-908.

[27] P. S. Chowdhury, S. Saha, A. Patra, Influence of nanoenvironment on luminescence of Eu^{3+} activated SnO_2 nanocrystals. *Solid State Commu.* 2004, 131, 785–788.

[28] F. Gu, S. F. Wang, M. K. Lu, Y. X. Qi , G, J, Zhou, D. Xu, D. R. Yuan, Luminescent characteristics of Eu^{3+} in SnO_2 nanoparticles. *Optical Mater.* 2004, 25, 59–64.

[29] X. Fua, H. Zhang, S. Niu, Q. Xin, Synthesis and luminescent properties of SnO_2:Eu nanopowder via polyacrylamide gel method. *J. Solid State Chem.* 2005, 178, 603–607.

[30] V. Mahalingam, M. Tan, P. Munusamy, Joe B. Gilroy, Frank C. J. M. van Veggel, Bright blue photo- and electroluminescence from Eu^{2+}-doped GaN/SiO_2 nanocomposites. *Adv. Funct. Mater.* 2007, 17, 3462–3469.

[31] G. A. H. Flores, Gallium nitride and GaN:Eu nanocrystalline luminescent powders. *Phys. Stat. Sol. A* 2008, 205, 43–46.

[32] C. Fu, J. Liao, W. Luo, R. Li, X. Chen, Emission of 1.53 μm originating from the lattice site of Er^{3+} ions incorporated in TiO_2 nanocrystals. *Opt. Lett.* 2008, 33, 953-955.

[33] X. Chen. W. Luo, Y. Liu, G. Liu, Recent Progress on spectroscopy of lanthanide ions incorporated in semiconductor nanocrystals. *J. Rare Earths* 2007, 25, 515-525.

[34] K. Riwotzki, M. Haase, Wet-chemical synthesis of doped colloidal nanoparticles: YVO_4:Ln (Ln = Eu, Sm, Dy). *J. Phys. Chem. B* 1998, 102, 10129-10135.

[35] B. Bihari, H. Eilers, B. Tissue, Spectra and dynamics of monoclinic Eu_2O_3 and $Eu^{3+}:Y_2O_3$ nanocrystals. *J. Lumin.* 1997, 75, 1-10.

[36] J. Bellessa, S. Rabaste, J. C. Plenet, J. Mugnier, O. Marty, Eu^{3+}-doped microcavities fabricated by sol–gel process. *Appl. Phys. Lett.* 2001, 79, 2142-2144.

[37] G. A. Hebbink, J. W. Stouwdam, D. N. Reinhoudt, C. Veggel, Lanthanide(III)-doped nanoparticles that emit in the near-infrared. *Adv. Mater.* 2002, 14, 1147-1150.

[38] D. K. Williams, B. Bihari, B. M. Tissue, Preparation and fluorescence spectroscopy of bulk monoclinic $Eu^{3+}:Y_2O_3$ and comparison to $Eu^{3+}:Y_2O_3$ nanocrystals. *J. Phys. Chem. B.* 1998, 102, 916-920.

[39] R. S. Meltzer, K. S. Hong, Electron-phonon interactions in insulating nanoparticles: Eu_2O_3. *Phys. Rev. B.* 2000, 61, 3396-3403.

[40] R. S. Meltzer, S. P. Feofilov, B. Tissue, Dependence of fluorescence lifetimes of $Y_2O_3:Eu^{3+}$ nanoparticles on the surrounding medium. *Phys. Rev. B* 1999, 60, R14012.

[41] D. K. Williams, H. Yuang, B. M. Tissue, Size dependence of the luminescence spectra and dynamics of $Eu^{3+}:Y_2O_3$ nanocrystals. *J. lumin.* 1999, 83–84, 297-300.

[42] H. Peng, H. Song, B. Chen, J. Wang, S. Lu, J. Zhang, Spectral difference between nanocrystalline and bulk $Y_2O_3:Eu^{3+}$. *Chem. Phys. Lett.* 2003, 370, 485-489.

[43] X. Bai, H. Song, L. Yu, Z. Liu, L. Yang, G. Pan, S. Lu, X. Ren, Y. Lei, L. Fan, Luminescent properties of pure cubic phase $Y_2O_3:Eu^{3+}$ nanotubes/nanowires prepared by a hydrothermal method. *J. Phys. Chem. B* 2005, 109, 15236−15242

[44] C. F. Wu, W. P. Qin, G. S. Qin, D. Zhao, J. S. Zhang, S. H. Huang, S. Lu, H. Liu, H, Lin, R. Lenne, Photoluminescence from surfactant-assembled $Y_2O_3:Eu$ nanotubes. *Appl. Phys. Lett.* 2003, 82, 520-522.

[45] Z. Wei, L. Sun, C. Liao, X. Jiang, C. Yan, Size-dependent chromaticity in $YBO_3:Eu$ nanocrystals: correlation with microstructure and site symmetry. *J. Phys. Chem. B* 2002, 106, 10610-10617.

[46] Z. Wei, L. Sun, C. Liao, C. Yan, Size dependence of luminescent properties for hexagonal YBO$_3$:Eu nanocrystals in the vacuum ultraviolet region. *App. Phys. Lett.* 2002, 80, 1447-1449.

[47] Z. Wei, L. Sun, C. Liao, X. Jiang, C. Yan, Y. Tao, X. Hou, X. Ju, Size dependence of luminescent properties for hexagonal YBO$_3$:Eu nanocrystals in the vacuum ultraviolet region. *J. Appl. Phys.* 2003, 93, 9783-9788.

[48] L. Yu, H. Song, Z. Liu, L. Yang, S. Lu, Fabrication and photoluminescent characteristics of La$_2$O$_3$: Eu^{3+} nanowires. *Physical Chemistry Chemical Physicas*, 2006, 8, 303-308.

[49] L. Yu, H. Song, S. Lu, Z. Liu, L. Yang, X. Kong, Luminescent properties of LaPO$_4$:Eu nanoparticles and nanowires. *J. Phys. Chem. B* 2004, 108, 16697-16702

[50] H. Song, L. Yu, S. Lu, T. Wang, Z. Liu, L. Yang, Remarkable differences of photoluminescent properties between LaPO$_4$:Eu one-dimensional nanowires and zero-dimensional nanoparticles. *Appl. Phys. Lett.* 2004, 85, 470-472.

[51] L. Yu, H. Song, S. Lu, Z. Liu, L. Yang, Influence of shape anisotropy on photoluminescence characteristics in LaPO$_4$:Eu nanowires. *Chem. Phys. Lett.* 2004, 399, 384-388.

[52] H. Song, J. Wang, B. Chen, H. Peng, S. Lu, Size-dependent electronic transition rates in cubic nanocrystalline europium doped yttria. *Chem.. Phys. Lett.* 2003, 376, 1-5.

[53] H. Peng, H. Song, B. Chen, J. Wang, S. Lu, X. Kong, J. Zhang, Temperature dependence of luminescent spectra and dynamics in nanocrystalline Y$_2$O$_3$: Eu^{3+}. *J. Chem. Phys.* 2003, 118, 3277-3282.

[54] W. P. Zhang, P. B. Xie, C. K. Duan, K. Yan, M. Yin, L. R. Lou, S. D. Xia, J. C. Krupa, Preparation and size effect on concentration quenching of nanocrystalline Y$_2$SiO$_5$:Eu. *Chem. Phys. Lett.* 1998, 292, 133-136.

[55] C. Jia, L. Sun, F. Luo, X. Jiang, L. Wei, C. Yan, Structural transformation induced improved luminescent properties for LaVO$_4$:Eu nanocrystals. *App. Phys. Lett.* 2004, 84, 5305-5307.

[56] L. Yu, H. Song, Z. Liu, L. Yang, S. Lu, Z. Zheng, Electronic transition and energy transfer process in LaPO$_4$-Ce^{3+}/Tb^{3+} nanowires. *J. Phys. Chem. B* 2005, 109, 11450-11455.

[57] L. Yu, H. Song, Z. Liu, . Yang, S. Lu, Z. Zheng, Remarkable improvement of brightness for the green emissions in Ce^{3+} and

Tb^{3+} co-activated LaPO$_4$ nanowires *Solid State Commu.* 2005, 134, 753—757.

[58] Q. Li, L. Gao, D. Yan, Effects of the coating process on nanocale Y$_2$O$_3$:Eu^{3+} powders. *Chem. Mater.* 1999, 11, 533-535.

[59] K. Kompe, H. Borchert, J. Storz, A. Lobo, S. Adam, T. Moller, M. Hasse, Green-emitting CePO$_4$:Tb/LaPO$_4$ core-shell nanoparticles with 70% photoluminescence quantum yield. *Angew. Chem. Int. Ed.* 2003, 42, 5513-5516.

[60] O. Lehmann, K. Kompe, M. Haase, Synthesis of Eu^{3+}-doped core and core/shell nanoparticles and direct spectroscopic identification of dopant sites at the surface and in the interior of the particles. *J. Am. Chem. Soc.* 2004, 126, 14935-14942.

[61] A. Huignard, V. Buissette, A. Franville, T. Gacoin, J. Boilot, Emission processes in YVO$_4$:Eu nanoparticles. *J. Phys. Chem. B* 2003, 107, 6754-6759.

[62] X. Bai, H. Song, G. pan, Z. Liu, S. Lu, W. Di, X. Ren, Y. Lei, Q. Dai, L. Fan, Luminescent enhancement in europium-doped yttria nanotubes coated with yttria. *Appl. Phys. Lett.* 2006, 88, 143104.

[63] L. R. Singh, R. S. Ningthoujam, V. Sudarsan, I. Srivastava, S. D. Singh, G. K. Dey, S. K. Kulshreshtha, Luminescence study on Eu^{3+} doped Y$_2$O$_3$ nanoparticles: particle size, concentration and core-shell formation effects. *Nanotech.* 2008, 19, 055201.

[64] M. Chang, S. Tie, Fabrication of novel luminor Y$_2$O$_3$:Eu^{3+}@SiO$_2$@YVO$_4$:Eu^{3+} with core/shell heteronanostructure. *Nanotech.* 2008, 19, 075711.

[65] G. Pan, H. Song, X. Bai, Z. Liu, H. Yu, W. Di, S. Li, L. Fan, X. Ren, S. Lu, Novel energy-transfer route and enhanced luminescent properies in YVO$_4$:Eu^{3+}/YBO$_3$:Eu^{3+} composite. *Chem. Mater.* 2006, 18, 4526-4532.

[66] M. Yu, J. Lin, J. Fang, Silica spheres coated with YVO$_4$:Eu^{3+} layers via sol-gel process: A simple method to obtain spherical core-shell phosphors. *Chem. Mater.* 2005, 17, 1783-1791.

[67] C. K. Lin, H. Wang, D. Kong, M. Yu, X. M. Liu, Z. L. Wang, J. Lin, Silica supported submicron SiO$_2$@Y$_2$SiO$_5$:Eu^{3+} and SiO$_2$@Y$_2$SiO$_5$:Ce^{3+}/Tb^{3+} spherical particles with a core-shell structures: Sol-gel synthesis and characterization. *Eur. J. Inorg. Chem.* 2006, 18, 3667-3675.

[68] C. K. Lin, D. Kong, X. M. Liu, H. Wang, M. Yu, J. Lin, Monodisperse and core-shell-structured $SiO_2@YBO_3:Eu^{3+}$ spherical particles: Synthesis and characterization. *Inorg. Chem.* 2007, 46, 2674-2681.

[69] H. Wang, C. K. Lin, X. M. Liu, J. Lin, M. Yu, Monodisperse spherical core-shell-structured phosphors obtained by functionalization of silica spheres with $Y_2O_3:Eu^{3+}$ layers for field emission displays. *Appl. Phys. Lett.* 2005, 87, 181907.

[70] G. Li, M. Yu, Z. wang, J. Lin, R. Wang, J. Fang, Sol-gel fabrication and photoluminescence properties of $SiO_2@Gd_2O_3:Eu^{3+}$ core-shell particles. *J. Nanosci. Nanotech.* 2006, 6, 1416-1422.

[71] P. Jia, X. Liu, M. Yu, Y. Luo, J. Fang, J. Lin, Luminescence properties of sol-gel derived spherical $SiO_2@Gd_2(WO_4)_3:Eu^{3+}$ particles core-shell structures. *Chem. Phys. Lett.* 2006, 424, 358-363.

[72] C. K. Lin, B. Zhao, Z. Wang, M. Yu, H. Wang, D. Kong, J. Lin, Spherical $SiO_2@GdPO_4:Eu^{3+}$ core-shell phosphors: sol-gel synthesis and characterization. *J. Nanosci. Nanotech.* 2007, 7, 542-548.

[73] H. Song, B. Chen, H. Peng, J. Zhang, Light-induced change of charge transfer band in nanocrystalline Y_2O_3: Eu^{3+}. *Appl. Phys. Lett.* 2002, 81, 1776-1778.

[74] Q. Lu, A. H. Li, F. Y. Guo, L. Sun, L. C. Zhao, The two-photon excitation of SiO_2-coated $Y_2O_3:Eu^{3+}$ nanoparticles by a near-infrared femtosecond laser, *Nnaotec.* 2008, 19, 1-8.

[75] P. A. Tanner, Synthesis and luminescence of nano-insulator doped with lanthanide ions. *J. Nanosci. Nanotech.* 2005, 5, 1455-1464.

[76] Y. Tao, G. Zhao, W. Zhang, S. Xia, Combustion synthesis and photoluminescence of nanocrystalline Y_2O_3:Eu phosphors. *Mater. Res. Bull.* 1997, 32, 501-506.

[77] M. Jia, J. Zhang, S. Lu, J. Sun, Y. Luo, X. Ren, H. Song, X. Wang, UV excitation properties of Eu^{3+} at the S_6 site in bulk and nanocrystalline cubic Y_2O_3. *Chem. Phys. Lett.* 2004, 384, 193-196.

[78] Z. Qi, C. Shi, W. Zhang, T. Hu, Local structure and luminescence of nanocrystalline Y_2O_3:Eu. *App. Phys. Lett.* 2002, 81, 2857-2859.

[79] T. Igarashi, M. Ihara, T. Kusunoki, K. Ohno, Relationship between optical properties and crystallinity of nanometer Y_2O_3:Eu phosphor. *App. Phys. Lett.* 2000, 76, 1549-1551.

[80] A. Konrad, T. Fries, A. Gahn, F. Kummer, U. Herr, R. Tidecks, K. Samwer, Chemical vapor synthesis and luminescence properties of nanocrystalline cubic Y_2O_3:Eu. *J. Appl. Phys.* 1999, 86, 3129-3133.

[81] Z. Fu, S. Zhou, Y. Yu, S. Zhang, Photoluminescence properties and analysis of spectral Structure of Eu^{3+}-doped SrY_2O_4. 2005, 109, 23320–23325.

[82] J. Wang, Y. Chang, H. Chang, S. Li, L. Huang, X. Kong, Local structure dependence of the charge transfer band in nanocrystalline Y_2O_3:Eu^{3+}. *Chem. Phys. Lett.* 2005, 405, 314-317.

[83] A. M. Pires, M. F. Santos, M. R. Davalos, E. B. Stucchi, The effect of Eu^{3+} ion doping concentration in Gd_2O_3 fine spherical particles. *J. Alloy Comps.* 2002, 344, 276-279.

[84] W. Strek, E. Zych, D. Hreniak, Doped polymers with Ln(III) complexes: simulation and control of light colors. *J. Alloy Comp.* 2002, 344, 320-322.

[85] R. Yan, Y. Li, Down/up conversion in Ln^{3+}-doped YF_3 nanocrystals. *Adv. Funct. Mater.* 2005, 15, 763-770.

[86] B. M. Tissue, H. B. Yuan, Structure, particle size, and annealing of gas phase-condensed Eu^{3+}:Y_2O_3 nanophosphors. *J. Solid. State. Chem.* 2003, 171, 12-18.

[87] J. C. Boyer, F. Vetrone, J. A. Capobianco, A. Speghigi, M. Bettinelli, Variation of fluorescence lifetimes and Judd-Ofelt parameters between Eu^{3+} boped Bulk and nanocrystalline cubic Lu_2O_3. *J. Phys. Chem. B* 2004, 108, 20137-20143.

[88] Q. Dai, H. Song, G. Pan, X. Bai, H. Zhang, R. Qin, L. Hu, H. Zhao, S. Lu, X. Ren, Surface defects and their influence on structural and photoluminescence properties in $CdWO_4$:Eu nanocrystals. *J. Appl. Phys.* 2007, 102, 054311.

[89] W. W. Zhang, W. P. Zhang, P. B. Xie, M. Yin, H. T. Chen, L. Jing, Y. S. Zhang, L. R. Lou, S. D. Xia, Optical properties of nanocrystalline Y_2O_3:Eu depending on its odd structure. *J. Colloid. Inter. Sci.* 2003, 262, 588-593.

[90] R. Bazzi, M. A. Flores-Gonzalez, C. Louis, K. Lebbou, C. Dujardin, A. Brenier, W. Zhang, O. Tillement, E. Bernstein, P. Perriat, Synthesis and luminescent properties of sub-5-nm lanthanide oxides nanoparticles. *J. Lumin.* 2003, 102–103, 445-450.

[91] G. Pan, H. Song, S. Lu, L. Yu, Z. Liu, X. Ren, X. Bai, Y. Lei, L. Fan, Structure and photoluminescent properties of microstructural YBO$_3$:Eu^{3+} nanocrystals. *J. Nanosci. Nanotech.* 2007, 7, 593-601.

[92] Q. Dai, H. Song, X. Bai, G. Pan, S. Lu, T. Wang, X. Ren, H. Zhao, Photoluminescence properties of ZnWO$_4$:Eu^{3+} nanocrystals prepared by a hydrothermal method. *J. Phys. Chem. C.* 2007, 111, 7586-7592.

[93] S. F. Wuister, C. de Mello Donega, A. Meijerink, Local-field effects on the spontaneous emission rate of CdTe and CdSe quantum dots in dielectric media. *J. Chem. Phys.* 2004, 121, 4310-4315.

[94] C. K. Duan, M. F. Reid, Macroscopic models for the radiative relaxation lifetime of luminescent centers embedded in surrounding media. *Spectr. Lett.* 2007, 40, 237-246.

[95] L. Yu, H. Song, S. Lu, Z. Liu, L. Yang, T. Wang, X. Kong, Thermal quenching characteristics in LaPO$_4$:Eu nanoparticles and nanowires. *Mater. Res. Bull.* 2004, 39, 2083–2088.

[96] H. Song, J. Wang, Dependence of photoluminescent properties of cubic Y$_2$O$_3$: Tb^{3+} nanocrystals on particle size and temperature. *J. Lumin.* 2006, 118, 220-226.

[97] H. P. You, M. Nogami, Three-photon-excited fluorescence of Al$_2$O$_3$-SiO$_2$ glass containing Eu^{3+} ions by femtosecond laser irradiation. *Appl. Phys. Lett.* 2004, 84, 2076-2078.

[98] S. Q. Xu, W. wang, S. F. Zhu, B. Zhu, J. R. Qiu, Highly efficient red, green and blue upconversion luminescence of Eu^{3+}/Tb^{3+}-codoped silicate by femtosecond laser irradiation. *Chem. Phys. Lett.* 2007, 442, 492-495.

[99] M. Mortier, A. Monteville, G. patriarche, G. Maze, F. Auzel, New progresses in transparent rare-earth doped glass-ceramics. *Opt. mater.* 2001, 16, 255-267.

[100] M. Mortiery, Between glass and crystal: glass-ceramics, a new way for optical materials. *Philosophi. Mag. B* 2002, 82, 745-753.

[101] F. Lahoz, I. R. Martin, U. R. Rodriguez-Mendoza, I. Iparraguirre, J. Azkargorta, A. Mendioroz, R. Balda, J. Fernandez, V. Lavin, Rare earths in nanocrystalline glass–ceramics. *Opt. Mater.* 2005, 27, 1762-1770.

[102] B. Zhu, S. M. Zhang, S. F. Zhou, N. Jiang, J. R. Qiu, Enhanced upconversion luminescence of transparent Eu^{3+}doped glass-

ceramics containing nonlinear optical microcrystals. *Opt. Lett.* 2007, 32, 653-655.
[103] Ho-Soon Yang, K. S. Hong, S. P. Feolov, Brian M. Tissue, R. S. Meltzer, W. M. Dennis, Electron-phonon interaction in rare earth doped nanocrystals. *J. Lumin.* 1999, 83-84, 139-145.
[104] S. P. Feofilov, Spectroscopy of dielectric nanocrystals doped by rare-earth and transition-metal. *Phys. Solid State* 2002, 44, 1407-1414.
[105] G. K. Liu, H. Z. Zhang, X. Y. Chen, Restricted phonon relaxation and anomalous thermalization of rare earth ions in nanocrystals. *Nano. Lett.* 2002, 2, 535-539.
[106] G. K. Liu, X. Y. Chen, H. Z. Zhuang, S. Li, R. S. Niedbala, Confinement of electron–phonon interaction on luminescence dynamics in nanophosphors of $Er^{3+}:Y_2O_2S$. *J. Solid. State. Chem.* 2003, 171, 123-132.
[107] B. Mercier, C. Dujardin, G. Ledoux, C. Louis, O. Tillement, P. Perriat, Confinement effects in sesquioxydes. *J. Lumin.* 2006, 119–120, 224-227.
[108] H. S. Yang, S. P. Feofilov, D. K. Williams, J. C. Milora, B. M. Tissue, R. S. Meltzer, W. M. Dennis, One phonon relaxation processes in $Y_2O_3 : Eu^{3+}$ nanocrystals. *Physica B: Conden. Matter.* 1999, 263–264, 476-478.
[109] J. W. Stouwdam, F. C. J. M. Van Veggel, Improvement in the luminescence properties and processability of LaF_3/Ln and $LaPO_4$/Ln nanoparticles by surface modification. *Langmuir* 2004, 20, 11763-11771.
[110] W. H. Di, X. J. Wang, B. J. Chen, S. Z. Lu, X. X. Zhao, Effect of OH^- on the luminescent efficiency and lifetime of Tb^{3+}-doped yttrium orthophosphate synthesized by solution precipitation. *J. Phys. Chem. B* 2005, 109, 13154-13158.
[111] D. Boyer, G. Bertrand, R. Mahiou, A spectroscopic study of the vaterite form $YBO_3:Eu^{3+}$processed by sol–gel technique. *J. Lumin.* 2003, 104, 229-237.
[112] G. Bertrand-Chadeyron, M. El-Ghozzi, D. Boyer, R. Mahiou, J. C. Cousseins, Orthoborates processed by soft routes: correlation luminescence structure. *J. Alloy. Compd.* 2001, 317, 183-185.
[113] M. Ren, J. Lin, Y. Dong, L. Yang, M. Su, Structure and phase transition of $GdBO_3$. *Chem. Mater.* 1999, 11, 1576-1580.

[114] E. M. Levin, C. R. Robbins, J. L. Warring, Immiscibility and the system lanthanum oxide–boric oxide. *J. Am. Ceram. Soc.* 1961, 44, 81-108.

[115] J. Zhang, J. Lin, Vaterite-type $YBO_3:Eu^{3+}$ crystals: hydrothermal synthesis, morphopogy, and photoluminescence properties. *J. Crys. Growth* 2004, 271, 207-215.

[116] G. E. Venikouas, R. C. Powell, Laser time-resolved spectroscopy investigation of energy transfer in Eu^{3+} and Er^{3+} doped YVO_4. *J. Lumin.* 1978, 16, 29-45.

[117] M. Yu, J. Lin, S. Wang, Effects of x R^{3+} on luminescent properties of Eu^{3+} in nanocrystalline $YVPO_4:Eu^{3+}$ and $RVO_4:Eu^{3+}$ thin film phosphors. *Appl. Phys. A* 2005, 80, 353-360.

[118] W. L. Wanmaker, A. Bril, J. W. ter Vrugt, J. Broos, Luminescent properties of Eu-activated phosphors of the type ABO_4. *Philips Res. Rep.* 1966, 21, 270-282.

[119] B. C. Chakoumakos, M. M. Abraham, L. A. Boatner, Crystal structure refinements of zircon-type MVO_4 (M=Sc, Ce, Pr, Nd, Tb, Ho, Er, Tm, Yb, Lu), *J. Solid State Chem.* 1994, 109, 197-202.

[120] W. Chen, Jan-Olle Malm, V. Zwiller, R. Wallenberg, Jan-Olov Bovin, Size dependence of Eu^{2+} fluorescence in $ZnS:Eu^{2+}$ nanoparticles. *J. Appl. Phys.* 2001, 89, 2671-2675.

INDEX

A

Abraham, 50
anisotropy, 16, 45
annealing, 48
applications, 2, 3, 24, 27, 31
atoms, 5, 8, 11, 19
authors, 3, 7, 11, 13

B

band gap, 37
bulk materials, vii, 4, 5, 7, 14, 18, 20, 24
burning, 3, 31, 39

C

chemical properties, 2, 5
chemical vapor deposition, 7
color, 1, 2, 34
combustion, 3, 7, 12, 19, 23
components, 1, 9
composites, 34
compounds, 1, 2, 3, 41
concentration, 4, 7, 9, 45, 46, 47
condensation, 12
configuration, 1, 2, 37
conversion, 3, 47
coordination, 7, 11
correlation, 44, 50
covalency, 2
crystalline, 3, 11, 13, 15, 27
crystallinity, 47
crystals, 14, 32, 42, 50

D

decay, 2, 4, 17, 33, 35
defects, 8, 9, 19, 27, 33, 48
density, 17, 27, 32
diode laser, 27
displacement, 8
distribution, 12, 23
dopants, 37
dynamics, 12, 39, 43, 44, 45, 49

E

earth, 1, 41, 42, 43, 49
electroluminescence, 43
electromagnetic, 37
electron, 3, 7, 23, 49
electrons, 8, 23
emission, 1, 2, 3, 4, 11, 12, 14, 15, 17, 21, 23, 25, 29, 33, 35, 36, 37, 38, 43, 46, 48

energy, 2, 3, 7, 8, 19, 23, 27, 28, 37, 43, 45, 46, 50
energy transfer, 3, 29, 43, 45, 50
environment, vii, 1, 3, 9, 11
ESR, 37
europium, 1, 39, 45, 46
evaporation, 3
excitation, 3, 4, 11, 12, 14, 15, 16, 23, 24, 25, 27, 28, 29, 33, 36, 38, 47

F

fluorescence, 3, 33, 35, 42, 44, 48, 50
fluorescence decay, 33, 35
formula, 24
functionalization, 5, 46

G

gel, 3, 43, 46, 50
generation, 29
glasses, 28, 29, 32
groups, 3, 9, 17, 32

H

host, 2, 3, 7, 17, 24, 28, 37, 42
hydrothermal synthesis, 3, 50
hydroxide, 3
hypersensitivity, 2, 11

I

ideal, 11, 34
identification, 45
interaction, 3, 32, 49
interactions, 44
interface, 27, 34
inversion, 1
ions, vii, 1, 2, 5, 7, 8, 9, 11, 13, 14, 19, 23, 24, 27, 28, 31, 32, 33, 37, 39, 41, 42, 43, 47, 48, 49

irradiation, vii, 5, 7, 8, 9, 10, 27, 29, 39, 48, 49

L

lanthanide, 1, 3, 32, 41, 43, 47, 48
lanthanum, 50
lasers, 27, 28, 42
lattice parameters, 9
lifetime, 3, 12, 17, 18, 19, 20, 21, 48, 50
ligand, 27
line, 1, 2, 3, 12, 28, 31, 32, 33
low temperatures, 2, 32
luminescence, vii, 3, 5, 12, 18, 23, 27, 37, 41, 42, 43, 44, 47, 49, 50
Luo, 42, 43, 45, 46, 47

M

manganese, 42
matrix, 28
media, 17, 18, 48
methanol, 42
microcrystalline, 29
micrometer, vii, 4, 14, 20
microstructure, 44
microstructures, vii, 9, 14
morphology, vii, 14

N

nanocomposites, 5, 33, 38, 43
nanocrystals, vii, 2, 4, 5, 7, 9, 11, 12, 14, 16, 17, 24, 27, 31, 33, 36, 37, 42, 43, 44, 45, 47, 48, 49
nanomaterials, vii, 14
nanometer, 14, 47
nanoparticles, vii, 2, 7, 9, 11, 12, 14, 15, 17, 19, 20, 21, 23, 24, 25, 27, 28, 31, 32, 33, 37, 38, 42, 43, 44, 45, 46, 47, 48, 49, 50
nanorods, vii
nanostructures, 27

nanowires, vii, 4, 14, 15, 16, 20, 21, 44, 45, 48
Netherlands, 41

O

optical properties, 43, 47
optoelectronics, 41
order, 2, 19, 28
oxides, 48

P

parameter, 24
parameters, 21, 24, 48
parity, 1, 37
particles, 8, 9, 11, 14, 19, 29, 37, 45, 46, 47
permission, iv, 8, 10, 12, 13, 14, 15, 16, 18, 19, 20, 21, 24, 25, 28, 29, 31, 35, 36, 38
phonons, 28, 32
photoluminescence, 1, 5, 37, 45, 46, 47, 48, 50
physics, 42
plasma, 34
polarization, 17, 20
polyacrylamide, 43
polymers, 47
population, 9, 23
power, 27, 28, 29
precipitation, 3, 50
probe, vii, 3, 5
properties, vii, 1, 2, 5, 16, 28, 34, 37, 39, 41, 42, 43, 44, 45, 46, 47, 48, 49, 50
pulse, 27
purity, 1, 34
pyrolysis, 3

Q

quantum confinement, 2, 19, 37
quantum dot, 48

quantum dots, 48

R

radiation, 17
radius, 2, 7
reactions, 3
recommendations, iv
refractive index, 3, 17
region, 37, 44
relaxation, 18, 19, 20, 34, 48, 49
relaxation process, 49
relaxation processes, 49
relaxation rate, 19
resolution, 3, 14, 16, 33
rods, 14

S

sapphire, 27
semiconductor, 2, 43
semiconductors, 2, 33
separation, 19, 23
shape, 9, 14, 16, 45
signals, 37
silica, 46
SiO_2, 5, 27, 28, 33, 37, 38, 43, 46, 47, 48
sol-gel, 2, 33, 46, 47
solid state, 3
space, 17
spectroscopy, 2, 3, 42, 43, 44, 50
spectrum, 34, 37, 38
Sun, 43, 44, 45, 47
surface modification, 5, 33, 49
surfactant, 44
symmetry, vii, 1, 3, 5, 9, 11, 12, 14, 15, 19, 34, 39, 44
synthesis, vii, 2, 7, 8, 9, 14, 43, 46, 47

T

temperature, 3, 12, 15, 18, 19, 20, 21, 23, 24, 31, 32, 48

temperature dependence, 31, 32
thermal quenching, 23, 24
thermalization, 49
threshold, 29
titania, 43
titanium, 27
transition, vii, 1, 2, 5, 7, 14, 17, 18, 19, 20, 23, 24, 27, 34, 37, 45, 49, 50
transition metal, 2
transition rate, 2, 5, 18, 19, 20, 24, 27, 45
transitions, 1, 2, 9, 11, 14, 15, 17, 21, 23, 29, 31, 32, 33, 37

U

UV irradiation, 8, 9
UV light, 9

V

vacuum, 11, 12, 17, 34, 44
vapor, 7, 47

W

wavelengths, 9, 14, 34

X

X-ray diffraction, 12
X-ray diffraction (XRD), 12

Y

yttrium, 50

Z

ZnO, 2, 4, 42
ZnO nanorods, 42
ZnO nanostructures, 42